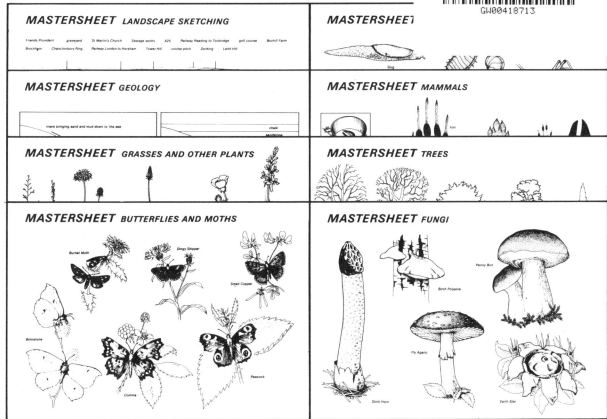

The mastersheets are at the back

INTRODUCTION

1. The thinking behind the pack

Box Hill has always been a popular destination for school trips and with the opening of the M25, the Hill has become even more accessible to Greater London. However today's teacher and pupil expect rather more from a day out than simply walking on the springy turf and enjoying the fresh air. Box Hill provides endless opportunities for experiencing and exploring the world outside the classroom. With this in mind, the Field Studies Council (pioneers of Environmental Education) and the National Trust (who own Box Hill) have produced this pack. It is designed for teachers of 7 - 15 year olds who are planning a school visit to the Hill. The pack includes background information, tried and tested ideas for investigations and mastersheets which will assist them in running a successful and valuable day for their pupils.

The contents of the pack were written by the teaching staff (Rachel Griffiths, Lucy Bonner, Pamela Johnson and John Bebbington) of the Field Studies Council's **JUNIPER HALL FIELD CENTRE,** incorporating ideas and material from the National Trust's Box Hill Manager and Warden. It was edited by the Warden and Director of Studies of Juniper Hall and the Box Hill Manager.

Juniper Hall Field Centre
by Vera Ibbett

The National Trust has a primary duty to conserve the natural history and amenities of Box Hill. All visitors are asked to observe the Country Code (as enclosed), the provisions of the Wildlife and Countryside Acts and the National Trust's bye-laws which prohibit the collection and removal of any plant or animal from its property.

2. Design and Contents of the Pack

The diagram shows how the pack covers topics at a number of scales. The mastersheets are at the back of the pack together with a map.

1
THE VIEW
2
GEOLOGY
3
HUMAN PATTERN
(Agriculture; Settlements; Communications)
4
SUCCESSION
(Downland; Scrubland; Woodland)
5
Grasses & Plants; Butterflies; Minibeasts;
Birds & Mammals; Trees; On the Ground.
6
PUBLIC PRESSURE
& CONSERVATION

3. Other information

Access to the Hill: The Hill lies between Dorking and Leatherhead off the A24. Pay and Display parking for cars is at the top of the Hill near the Information Centre. Coaches should not use the Zig Zag road as a weight restriction applies. They should either approach from the East (via the B2032 or the B2033) or park in the large coach park at the bottom of the Hill, opposite the Burford Bridge Hotel. The nearest train stations are within walking distance. The Boxhill and Westhumble station is on the London-Horsham line and Deepdene station is on the Tonbridge-Reading line. The 714 Green Line London-Horsham bus stops near the Burford Bridge Hotel (see inside front cover).

For individual permission, information regarding guided walks or opening times of the Information Centre, please contact the N.T. wardening staff on the following phone numbers:-

Dorking (0306) 885502 (office hours)
Dorking (0306) 884371
Dorking (0306) 887890

Facilities on the Hill: These are grouped together on the summit and include a shop (which sells a wide range of books and leaflets), toilets, refreshment facilities and an Information/Exhibition Centre which co-ordinates guided walks and a growing schools' liaison programme.

THE VIEW

A field course should normally start with a scene-setting session so that the pupils appreciate where more detailed studies fit in. Start with a simple question – What can you see? The fact that certain features are directly in front of the pupils' eyes does not mean that they will see them. By asking pupils to undertake a simple landscape sketch the teacher can find out what they are 'seeing'.

Put the view into perspective:
At the viewpoint (Salomons Memorial – erected in memory of Leopold Salomons who gave the National Trust its first land on Box Hill in 1914) distances (in miles) are given to various towns and features in the view. Although Box Hill is only 172m high, by looking due South on a clear day you will be able to see right across the Central Weald (see 'Geology') to Chanctonbury Ring on the South Downs, 25 miles away. It is an Iron Age hill fort and appears as a clump of trees on the far horizon.

Further to the left (South-east) the view is of the open Weald country: a patchwork of woods, fields and hedges. This was once a vast oak forest (see 'Human Pattern'). Watch out for aeroplanes taking off from and landing at Gatwick.

Below and to the right (South-west) is the town of Dorking and beyond it the wooded range of the Lower Greensand hills roughly parallel to the North Downs. The summit of Leith Hill – marked by a tower on the skyline–is almost 294m (1000ft.) and is the highest point in South-east England.

From just below the top of the Long Spur there are also impressive views due West across the Mole Valley to Ranmore – look for the prominent church spire.

Fieldwork and follow up ideas.

1. Landscape sketching.
Difficulties are often experienced in introducing pupils to sketching in the field and when attempting it for the first time the results can be disappointing. There are a couple of ways of overcoming this situation.
    (a) Taking a photograph of the view from which the general outline can be traced and duplicated for the children. The children then fill in the detail when they are out in the field, actually looking at the view.

(b) Using a sketching frame to "control" the sketch. It is basically a cardboard (or other) frame which hides the features outside the desired view. It may also help to have a grid of threads within it and matching sketching paper divided into similar squares.

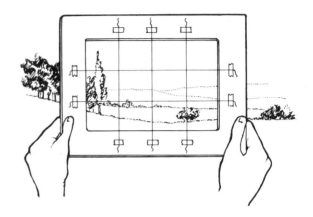

A sketching frame

(c) An easier way for younger pupils to appreciate the view and its component parts is to use the mastersheet. Colour (use different shades of one colour – dark for the foreground and lighter for the distance) and cut out the shapes of the hills, etc. Stick them onto each other in the appropriate place to see how they "fit" together.

2. Small groups can be asked to draw different parts of the view – these may then be joined together to form a panorama.

3. The advantages offered by the relatively high vantage point of the Hill can also be used in the study of maps. The view faces due South, which means that maps are 'upside-down' and this is a good opportunity to introduce map-and-compass work.

4. Landscape sketching offers opportunities for pupils' observations to lead to considerations and investigations of such things as the quality of their own immediate environment back at school and subsequent comparisons.

References:
Landscape drawing by G. Hutchings (Methuen).

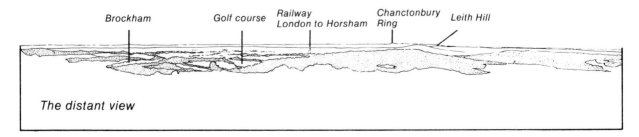

The distant view

Brockham Golf course Railway London to Horsham Chanctonbury Ring Leith Hill

GEOLOGY

The shape of a particular landscape is determined by a number of factors, the most important being the type of rock. There are three main types of rock:-

1. **Igneous rocks** are formed from hot molten material below the Earth's surface which then cools and solidifies (e.g. Lava and Granite).

2. **Sedimentary rocks** are formed from the particles derived from the wearing away or erosion of previously existing rocks, the remains of plants and animals or chemical deposits. They were laid down as sediments, usually in the sea, and later compressed to form solid rocks (e.g. Clay, Sandstone, Limestone, Shale and Chalk). This cycle of erosion and deposition is going on all the time, so 'new' rocks are being formed under the sea today from the sand and silt brought down by rivers.

3. **Metamorphic rocks** started out as either sedimentary or igneous rocks but have since been changed by intense heat or pressure. They are often very hard (e.g. Slate which is metamorphosed mud).

Box Hill (one of the best known summits in South-east England) and the North Downs (the chalk ridge which runs from Hampshire to Dover) are in an area called the Weald – strictly speaking this is the area **in between** the North and South Downs, but is sometimes taken to include them. The most distinctive feature of Box Hill is its position at the corner of one of the few 'gaps' in the line of the North Downs. Through this gap flows the River Mole (one of the rivers which drains the Weald), on its way Northwards to the Thames at Molesey (from where it gets its name). It flows through the Mole Gap, a valley which it has carved out over millions of years. Part of the steep cliff is known as 'The Whites' from the bare chalk which shows between the bushes and the trees.

When considering the formation of the Downs and The Weald, it is important to remember that Britain as we know it today originally lay a long way South and only later moved North to its present position by a process called Continental Drift, the slow movement of continents around the globe on a series of large 'plates'. It is at the boundaries of these plates that the greatest activity, in the form of volcanoes and earthquakes,

occurs e.g. the 'Ring of Fire' around the Pacific. The time-scale in the following formation may be difficult for the pupils to appreciate (see Fieldwork and follow up ideas).

200 million years ago: The area was a hard rocky plateau. This rock, called the Purbeck Beds, is the oldest rock to 'outcrop' in the Weald.

140-120 million years ago: The plateau was a swampy area and was gradually covered by sand and mud brought down by rivers. They became rocks and are called the Hastings Beds and the Weald Clay.

120 million years ago: The area was covered by a shallow sea and more sand and mud were deposited; this became rock known as the Lower Greensand.

110-100 million years ago: The sea deepened and more mud and sand were deposited. These became rocks known as the Gault Clay and the Upper Greensand.

100 million years ago: An easy date to remember. The area was in the warm sub-tropics and was covered by an extensive shallow sea. It was teeming with marine life, mainly tiny planktonic algae. These simple plants used the lime in the sea water to make their cell walls stronger, creating tiny skeleton-like structures known as 'Coccoliths'. When they died the skeletons all dropped to the sea floor. This carried on for 35 million years and gradually formed a thick layer of Chalk.

65 million years ago: 'Britain' started to move North.

25 million years ago: 'Britain' reached its present position. All the countries were moving about and Africa was pushing against Europe. This forced the land to buckle up into mountains (the Alps). The Weald experienced the outer ripples of this movement and was pushed into a wide shallow dome (called an anticline).

2 million years ago until the present: There has been massive erosion by rivers (e.g. the Mole) and periglacial processes (found in areas around the edge of the ice caps). The weakest point of an anticline is the middle (bend a long eraser to see this) and therefore it was the middle of the Weald which was eroded most. The Chalk and Lower Greensand were left as hills. The top of the chalk hills are capped with an 'icing' of clay-with-flints.

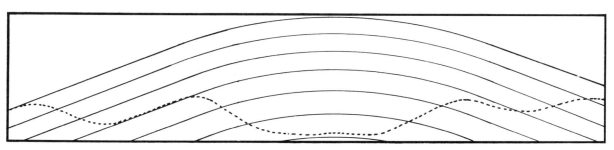

The Weald was pushed into an anticline and then eroded

Near Box Hill the Chalk is exposed on a large scale in a number of quarries. Chalk has been quarried at least since Roman times. It was spread on heavy clay soil to 'sweeten it', because putting chalk onto acid soil raises the pH and improves fertility. Hard bands of chalk were quarried for building, sometimes as single blocks and sometimes in combination with flint, which occurs as hard nodules in layers in the chalk. Chalk is still burned to provide mortar and agricultural lime and is also used on a massive scale in the manufacture of cement.

Properties of Chalk

Chalk is an Organic Sedimentary rock: it was laid down as a sediment derived from living organisms and often contains fossils of them. It is made of Calcium Carbonate ($CaCO_3$) which dissolves in acid. Chalk is both porous and permeable. It lets surface water drain through, therefore although it is a 'soft' rock, it is resistant to water erosion (under normal conditions) and is left upstanding as hills. Typical Chalk hills are smooth, rounded, covered in a thin soil and have a characteristic grassy vegetation resulting from centuries of sheep grazing.

Fossils

Fossils are the remains of animals and plants which lived on the Earth's surface millions of years ago. They are usually the hard parts of organisms which have fallen to the bed of the sea, lake or river and have subsequently been buried by the sediment accumulating there. As the sediment was turned to stone their remains became chemically altered so that their shapes have been preserved although the original matter has disappeared. The best place to look for fossils is in sedimentary rock. Fossils can be found in many states of preservation. For example, in shells, the soft part of the animal will have gone but the shell might be more or less in its original condition but very rarely complete. The original material of an organism is often dissolved away just leaving a hollow in the stone in the shape of the organism – this is called a 'mould'. The hollow mould may later be filled in by minerals. The resulting 'cast' has the actual shape of the original organism but NOT its structure or substance. 'Trace fossils' are preserved footprints or burrows and simply appear as marks on the bedding planes. Occasionally fossils can be so abundant in rock that they constitute the main part of the rock itself.

Fieldwork and follow up ideas.

1.   Are the slopes smooth and rounded? Survey the scarp slope using clinometers (bought or home-made) and ranging poles or people of a similar height. Draw to scale on graph paper and see if the slope has a 'typical chalk' shape (convex at the top, concave at the bottom with a straighter section in the middle). Compare with the dipslope or a slope on another rock type near school.

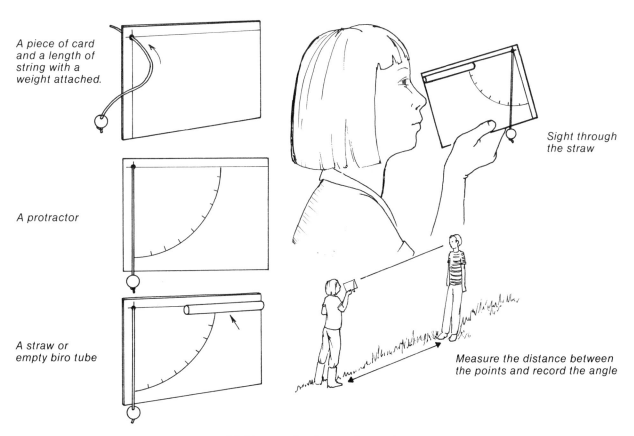

A piece of card and a length of string with a weight attached.

A protractor

A straw or empty biro tube

Sight through the straw

Measure the distance between the points and record the angle

Making and using a clinometer

2. Investigate the chemical and physical properties of chalk.

(a) pH test: grind up a small lump of chalk and then test its pH (see 'On the Ground'). The pH will be over 8, depending on length of time exposed to acids in the atmosphere, indicating strong alkalinity.

(b) $CaCO_3$ 'fizzy' test: drop a little acid (lemon juice or vinegar will do) onto a small piece of Chalk. Look and listen for a vigorous noisy frothing indicating a high $CaCO_3$ content.

(c) Hardness: Scratch with a penknife and compare with other rock types. On a hardness scale of 1 - 10; = Talc, 3 = Chalk, 10 = Diamonds.

3. Appreciation of the time-scale.
Life on earth began between 3,000 and 4,000 million years ago. This time can be represented as a 24 hours day: 1 hour = 150 million years, 1 second = 40,000 years. At what time did everything happen? At what time did man appear? A more visual representation is to use a piece of string 36 metres long. Each metre = 100 million years. Man appeared in the last mm.

4. Build contour models of Box Hill by tracing and cutting out successive contour lines onto polystyrene tiles. Stick them together, round off the edges, paint, cover with vegetation etc.

5. Make your own fossils using Plaster of Paris (available from hardware shops) and vinamoulds (rubbery moulds – the materials for making these are available from good model shops).

6. Investigate modern uses of chalk – how is cement made, where is it made? What is the difference between school 'chalk' and the real thing?

References:
The Nature Trail Book of rocks and fossils by M. Bramwell (Usborne).
The Weald by W. Gibbons (Unwin).
The Weald by S. Wooldridge & F. Goldring (Collins).

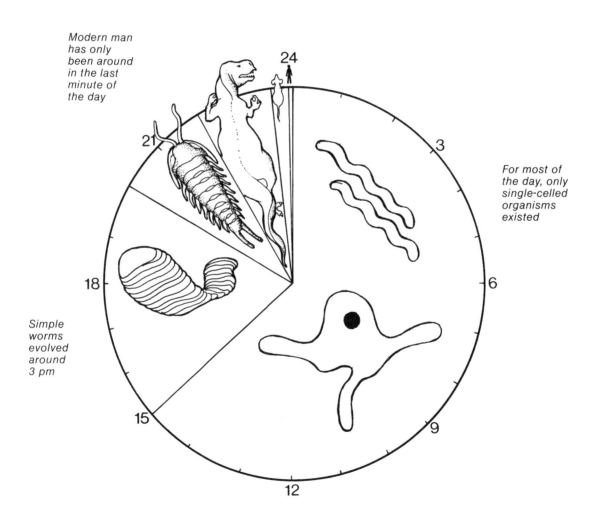

Modern man has only been around in the last minute of the day

For most of the day, only single-celled organisms existed

Simple worms evolved around 3 pm

The history of life in 24 hours

HUMAN PATTERN

Natural processes (see 'Geology') only create the general lie of the land. Upon these slopes and valleys man leaves his imprint with his demand for food, shelter and travel. It is important for children to realise how artificial the British landscape is (see 'Succession'). Get them to shut their eyes and listen for the rumble of cars and trains backed by the distant hum of human activity.

Farming came to Britain about 5000 years ago. From a hunting and gathering existence, man started to increase the productivity of certain plants and animals to obtain food and raw materials more easily. With limited tools and simple ploughs it was only possible to work lighter, chalky or sandy soil. The trees were cleared to make way for cultivation or grazing and the wood was used for fuel and to meet the demand for building materials, now that a sedentary way of life had been adopted. The heavy Weald Clay soils posed more of a problem and the dense woodland here was not cleared until after the Norman Conquest. Temporary settlements in the forest, originally used for iron extraction, timber felling or keeping pigs, developed into permanent villages to accommodate the increasing population.

The development of agriculture and settlements went hand in hand with an increase in trade. Self-sufficiency was no longer the norm and produce was exchanged and sometimes exported. This necessitated the development of communications. The oldest routes are the two ancient trackways running parallel with each other along the North Downs, connecting Wessex with the ports on the Kent coast. The higher of the two, the Ridgeway, afforded wide views which enabled the traveller to keep a look out for ambushes. The lower route with fewer "ups and downs" had obvious advantages in more settled times. The route along the North Downs is often called the Pilgrims' Way but had been in existence for centuries before any mediaeval pilgrims might have used this track on the way to Canterbury. Today running East–West, the A25 carries motor vehicles (although it has recently been relieved of much traffic with the opening of the M25). The railway, opened in 1849 by South Eastern Railways, follows a similar line from Guildford to Redhill.

The other important routeway runs North–South. The line of the Downs forms a barrier to traffic travelling this way between London and the South Coast. The River Mole, draining from the Weald into the Thames, cuts through this barrier by carving a four-mile gap through the chalk between Dorking and Leatherhead. Box Hill forms one corner of this gap. Although the Romans did not build Stane Street to run directly through this gap, other routeways have always seemed to use the valley. The prospect of the flat valley floor was much less daunting than tackling the scarp slopes, except in times of flood. When the Mole was running high, the traveller was forced to take the alternative road up and over the Downs. This was known as the Winter Road since it was obviously used more in the season of highest rainfall. Today running North–South, the A24 connects London with the South Coast. The dual-carriageway by-pass of Mickleham was built in the 1930s. The railway line up to London was opened in 1867 by London, Brighton and South Coast Railways. The charming Boxhill and Westhumble station was used by many day trippers from South London visiting the Hill.

From the viewpoint, it is possible to see a third mode of transport as planes take off from and land at Gatwick airport, the fourth busiest international airport in the world. About 15 million passengers per annum use the airport. On peak days, 500 flights take off and land on a single runway 3098 metres long and 46 metres wide – this can mean as many as 50 'movements' an hour. A second terminal is due to open in 1987 which will further increase the airport's capacity to 25 million passengers per annum.

The prosperity of Crawley, a new town, has been credited to the success of Gatwick and it is true to say that communications have always stimulated the development of settlements.

One settlement, that of Dorking, can be seen from the viewpoint. The 65m spire of St. Martin's church provides a distinctive feature in the landscape. The sewage works are also noticeable, sometimes to both eyes and nose especially when the sewage is pumped out onto the fields of Lower Boxhill Farm. The huge nettles thrive on the high nitrogen and phosphate levels. It is no coincidence that Dorking is situated where the East – West and North – South routes cross: by being at this node in communications it was able to develop as a market town for the surrounding villages. On market day livestock was sold in what is now the High Street, while the farmers' wives were thought to have sold dairy products in Butter Hill off South Street.

The communications also provided trade for a good many inns, equivalent to today's 'Happy

Eaters' and the like, where travellers could feed and rest themselves and their horses. The rather grand Burford Bridge Hotel used to be a small roadside inn, which could boast Nelson among its guests, who is reputed to have stayed here. Before the main road by-passed Mickleham, the Running Horses in the village was an important stopping place, while in Dorking many inns were to be found, of which the White Horse is the finest remaining example.

The road and rail links determine much of the area's character today. Strengthened further by the opening of the M25, they provide a quick and easy journey into London for many of the local residents who commute daily into the city. Offices also have taken advantage of the communications, and the green roofs of the 'Friends Provident' buildings are easily seen from the viewpoint. Both residents and workers enjoy the country-side of Box Hill and its surroundings which are protected by 'green belt' planning policy which was brought in after the Second World War and implemented strict control on most industrial and residential development in an effort to curtail the London sprawl.

Fieldwork and follow up ideas.

1. The impact of routeways. The importance of routeways to settlement development can be shown when landscape sketching (see 'The View').

2. Comparing routeways. From vantage points (the viewpoint and Burford Slopes) count the number of cars and trains travelling along the East – West and the North – South routes in the same time period. Additional train data can be obtained from timetables. Almost all the trains which use these lines are passenger services. Present the results as bar charts. A more sophisticated survey can distinguish between cars and lorries etc. travelling on the roads.

SUCCESSION

Bare rock will not remain bare for long if climatic conditions are conducive to plant growth. Even without soil present lichens can colonise it, obtaining all necessary nutrients from rainwater and the breakdown of the rock. Following the death of these **'pioneer'** plants, organic matter will start to accumulate, not only providing nutrients for further plant growth but also increasing chemical weathering of the rock and causing airborne particles to be trapped. This describes the initial formation of soil.

Once a depth of soil has been formed grasses and herbaceous plants will be able to establish themselves. More organic matter will collect as these plants die, giving yet more depth to the soil and altering conditions so much that the original pioneer plants may not survive at all. In this way one type of plant community **'succeeds'** another. The animals associated with each plant community will change too, so that once shrubs have colonised the area then mammals and birds may begin to use the area for a home. Shrubs require a deeper soil in which to flourish and once established they can add vast quantities of leaf litter to the soil. Ultimately the area will become woodland.

On Box Hill, the woodland on the chalk is dominated by beech, with yew as the understorey; on the more acid clays, oak and ash are more successful. A stable community can create an environment which is favourable to its own survival and may exclude other species. This stable community is called a **'climax'** community and its make-up depends on the controls of climate and soils. In Britain's cool temperate climate, oak woodland usually represents the climax community but on the very alkaline soils of the chalk, beech is more successful. Where the soil is shallower and the gradient steeper, yew may become the climax, as on The Whites.

However it is very clear that the climax community does not exist in most of Surrey. Man's activities have checked the sequence in a number of ways. In some cases it is completely replaced by arable farming, golf courses, roads, buildings etc. (see ('Human Pattern'). In others, succession is stopped at a certain level e.g. by grazing animals so that grassland prevails. This occurs because grasses have basal meristems (they grow from the base) and when the grass blades are bitten off the plant can continue to grow from ground level. Conversely trees and shrubs have apical meristems (they grow from the tip) and once the tip has been bitten off they can no longer grow. Halting the successional sequence is not necessarily a bad thing; it creates a diverse range of habitats such as pockets of woodland, tangles of scrubland and larger areas of grassland; the characteristic 'patchwork quilt' which animals and plants exploit.

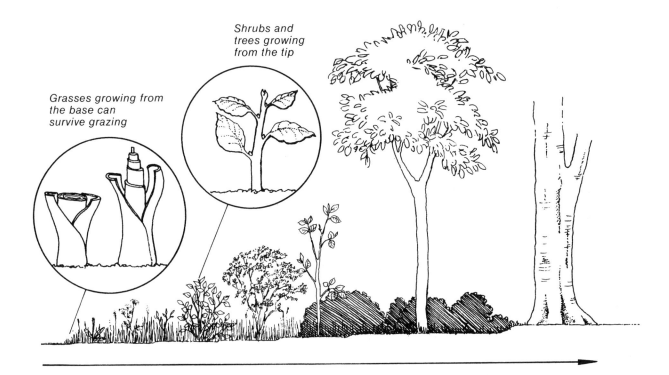

Grasses growing from the base can survive grazing

Shrubs and trees growing from the tip

With time, one type of plant community 'succeeds' another

Downland can be defined as the almost treeless, undulating chalk uplands used for grazing. It is the grazing which is crucial, since it prevents the climax community of woodland developing. The downland areas of Box Hill are maintained by a programme of periodic sheep grazing and are on the slopes of the Zig-Zag valley, the Long Spur, the South-facing slope below the viewpoint and Juniper Top. Because the vegetation is short, there is only a small degree of shading, so a great variety of plant species, including many attractive wild flowers, have established themselves on the downland. As a result of this great plant diversity, there is also great animal diversity (see 'Minibeasts' and also 'Butterflies and Moths').

"Scrub" is a term which is widely used and easy to picture in the mind's eye, but rarely defined. It is basically an area 'dominated by shrubs and bushes' and is taken to be the overlapping stage between the preceding grassland and subsequent woodland stages. It is therefore almost impossible to say when the scrub phase has begun or ended. It is rich in all forms of animal life; many species are specific to this type of habitat. A great problem is to maintain this transient stage indefinitely.

A section on woodlands logically follows one on scrubland. However very few of the trees on Box Hill are more than 180-200 years old and much of the woodland is clearly of much more recent origin; for example, there are a number of conifer plantations which were established in the late 1940s. All the woods provide many habitats for birds, mammals etc. To maintain this landscape man's activities have to continue otherwise it will revert to a tangle of scrub and eventually woodland.

With an increasing population striving for a better standard of living, man's impact on the landscape changes to meet these new demands. Field sizes increase to enable the use of bigger, more efficient machinery and it may no longer be economic to graze sheep on the Downs. Such changes have a profound effect on the landscape. Valuable hedge habitats are lost from the fields whilst on the Downs nature takes its course with succession leading to a shrub community preceding that of woodland and therefore management of the Downs has to be 'active' (see 'Public Pressure and Conservation').

Fieldwork and follow up ideas.

1. Can you recognise areas on Box Hill at different stages of the successional sequence? Study a transect from open downland into woodland. At 2 metre intervals along a straight line (of say 24 metres), measure the height of the three tallest plants that you can touch within an arm's length from the tape (for measuring the tall plants, see 'Trees'). Do not place the line straight along a path. Draw up the results as vertical straight lines to scale along a horizontal line equivalent to 24 metres long. Is there any evidence of downland turning to woodland via scrub?

2. What would happen to the areas of downland if grazing by sheep and rabbits stopped?

3. On a field sketch from the viewpoint, mark the different land uses which can be seen.

4. Find out about the National Trust's policies towards areas of woodland, scrubland and grassland. – Contact the N.T. wardening staff.

GRASSES AND OTHER PLANTS

On the downland slopes of Box Hill the underlying rock gives the soil an alkaline quality; therefore species tolerant of, or requiring calcium carbonate flourish. These are known as 'calcicoles'. The dominant grasses vary from place to place, but the most commonly found species will be sheep's fescue, tor grass and upright brome grass. The proportion of each grass species is partly determined by the trampling pressure on a patch of grassland (see Fieldwork and follow up ideas). The commoner herbaceous plants include daisy, salad burnet, mouse-eared hawkweed, marjoram, stemless thistle, St. John's-wort, small scabious and several others characteristic of chalk grassland. Some orchids are evident for a short period from May to July and there is a well developed moss layer beneath and amongst the herbaceous plants. Shrubs that may establish themselves include dogwood, hawthorn, and wayfaring tree. These areas of scrub provide a very useful habitat. In a recent Nature Conservancy Council (NCC) survey 152 species of invertebrates were found in a 30 x 20m area of isolated scrub.

The vegetation growing on the chalky soil (see 'On the Ground') has to be **adapted to dry conditions.** Several plants display methods of reducing water loss. Sheep's fescue has inrolled leaf blades which protect the tiny pores called 'stomata' from excessive water loss, hairiness seems to have the same effect, seen in mouse-eared hawkweed. Some plants such as stemless thistle have well-developed waxy cuticles and others such as common wild thyme have reduced leaf surface area. Other plants such as small scabious counteract water loss by producing long roots that penetrate the chalk beneath the soil or by developing water storage organs, e.g. bulbous buttercup. Salad burnet (whose leaves have a cucumber flavour) possesses deep roots that are well developed. This plant is perhaps the commonest species occuring on the downland of Box Hill, if grasses are excluded.

To survive **trampling** by the thousands of visitors to these grassy slopes, a number of varied adaptations have occurred. Hoary plantain shows an example of the rosette growth-habit. This flat way of growing allows the plant to avoid being kicked off by passing feet or being eaten by grazing animals; it also serves to reduce the drying effect of wind. The bedstraws display a mat growth habit where the stems grow laterally along the ground, again keeping the leaves and flowers out of the wind and away from grazing jaws or heavy boots. Mosses not only grow in a mat habit, some can also tolerate a large reduction in their water content if conditions become dry; if compressed underfoot they spring back to their original shape almost immediately the pressure is released. Underground storage organs enable a number of orchids to survive e.g. fragrant and common spotted. These are very vulnerable to being broken off whilst in flower but they persist because food made by the leaves is stored underground for the next growing season.

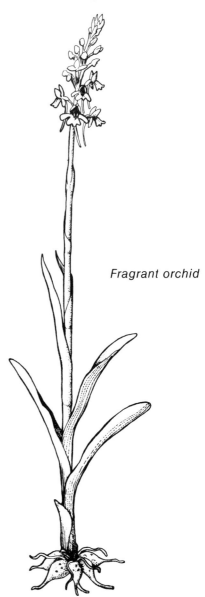

Fragrant orchid

Common thyme

Fieldwork and follow up ideas.

1.   Trampling survey. To measure the effect of feet, lay a piece of string or a tape so that it crosses a path, with the middle of the tape in the middle of the path. You now have a transect – a line along which you can sample at intervals. At the simplest level this can be just a case of measuring plant height with a ruler and soil depth with a long tent peg or strong knitting needle (with a cork on the end for safety in transit). With frames or quadrats, made from wood, string, coathangers or even cardboard in dry weather, more detailed investigations can be carried out. The number of different species within the quadrat or the presence/absence of certain species can be recorded by anybody who is prepared to get down close. Look for indicators of trampling e.g. plantain, daisy, fine-leaved grasses etc. The results can be presented on bar graphs (see 'Public Pressure and Conservation' for other ideas).

2.   Many chalkland plants are the food plants of butterflies and a study can be done relating the two (see 'Butterflies and Moths').

3.   Every school field is regularly 'trampled'. Look for plants that grow well and in abundance; why do they survive better than others?

4.   Study methods of reproduction by species that survive trampling. Does this information help explain how they survive?

References:
Wild flowers of the Chalk and Limestone by E. Ellis (Jarrold).
British wild orchids by H. Angel (Jarrold).
Concise British flora in colour by W. Keble Martin (Michael Joseph).
The Wild Flower key of the British Isles by F. Rose (Warne).

BUTTERFLIES AND MOTHS

Over 40 species of butterfly occur or have occurred on Box Hill. Six of these, including the Red Admiral, Painted Lady and Clouded Yellow are migrants from Southern Europe and could be seen anywhere, at any time between April and October.

Of the other species, the Silver Spotted Skipper, Duke of Burgundy and Adonis Blue, are rare, and not dealt with here.

Although butterflies can move quickly over quite long distances, most species tend to be confined to a particular habitat where they will visit favourite flowers and find foodplants for their caterpillars. Taking the two main habitats in turn – namely (a) Chalk grassland and (b) Woodland (not forgetting the important scrubland edge to the woodland) we can separate out two main groups of butterflies, both of which include species which are easy to recognise and interesting to observe.

(a) Chalk grassland
From late April to August a number of grassland species can be observed, including the Dingy Skipper – this is the earliest of the small butterflies to emerge and is a dull grey-brown, fast-flying and difficult to spot. It is the only British butterfly to settle with its wings folded forwards like a moth. When at rest it wraps itself around a dead flowerhead, e.g. knapweed. The caterpillar feeds on bird's-foot trefoil. Overlapping its season is

Dingy Skipper

the smaller Grizzled Skipper which is strongly marked in black and white. The caterpillar feeds on wild strawberry. It is interesting to watch the butterfly feeding on bugle flowers; its tongue is so short that it has to stand head down in order to reach the nectar (bugle flowers have no top lip, so access is easier this way).

In late May and June and again in August three of the common 'Blues' appear – the Common Blue, found throughout Britain, whose caterpillars feed on bird's-foot trefoil; the Small Copper whose caterpillars feed on common sorrel and the Brown Argus whose caterpillars feed on rock-rose. The first two species are jewel-like in their colours (which are produced by interference with light – the scales are virtually colourless). The scarcer Little or Small Blue – our smallest butterfly, which is dark brown above and grey beneath – can also be seen flying around patches of horseshoe vetch. It has a strange life cycle; eggs are laid on kidney vetch flowerheads in June, and the caterpillar has pupated by the end of July, remaining as a pupa for 10 months or more.

Small Copper

In June and July three of the larger species appear, the abundant Meadow Brown, the chequered Marbled White and the dashing Dark Green Fritillary. The Meadow Brown whose caterpillars eat soft grasses, has a female which is much brighter than the male. In fact the male's colours are hidden by a band of sooty black scent-scales which are used to persuade the female to mate.

Closely related to the Meadow Brown, the Marbled White is a striking black-and-white chequered butterfly which is common all over Box Hill. The caterpillars, like that of the Meadow Brown, feed on grasses. The Dark Green Fritillary is a striking tawny-orange butterfly which flies very quickly over open downland; the caterpillars feed on violets. There are three common and very similar Skipper butterflies on the wing in the Summer, first the orange-and-black Large Skipper, then the easily confused Small and Essex Skippers. These latter two can only be identified by examining the tip of the underside of the antenna which is orange in the Small Skipper and black in the Essex Skipper. No attempt should be made to catch and identify them, they can simply be listed as 'Small/Essex Skippers'.

Five and six-spot Burnet Moths, which fly in June, July and August are often mistaken for butterflies – they fly by day and are brilliantly coloured, but are in fact day-flying moths. Their bright red forewing spots and underwings advertise their distasteful nature – they contain cyanide compounds.

Burnet Moth

(b) Woodlands and scrub

Early in the year the Brimstone and the Comma may be seen feeding on sallow catkins, with the occasional Peacock and Small Tortoiseshell. The female Brimstone is creamy white, while the male is bright primrose in colour. In areas of chalk the eggs are laid on purging buckthorn. The caterpillars are carnivorous; larger ones eat smaller ones so that eventually there is only one large caterpillar per branch. The caterpillars can be found in May and June. The Comma is marvellously shaped like a ragged dead leaf and spends the Winter hibernating out in the open. The name comes from the white 'C' marking on the underside of the hindwing. Before hibernating the adult can be found feeding drunkenly on over-ripe blackberries. Eggs are laid in the Spring and again in the Summer on elm, hop and nettle.

In April, May and June the Green Hairstreak flies around sloe or gorse bushes; the butterfly, although one of the 'blues' is brown above and green below. The Speckled Wood, a member of the Brown family inhabits clearings and rides, where the caterpillars eat soft grasses such as Yorkshire fog. This dappled butterfly is well concealed in light-and-shade patterns. The White Admiral can be found where honeysuckle, the caterpillar's foodplant and bramble, the butterfly's favourite flower, grow together in woodlands. The butterfly has a stately gliding flight.

Finally the two largest butterflies on the Hill, both of which are hard to find, but both of which are spectacular. The Silver-Washed Fritillary is tawny orange, very swift on the wing and inhabits woodland rides and glades, feeding on bramble flowers. The underside is beautifully patterned in green and silver. The egg is laid on bark in August and the caterpillar hibernates throughout the Winter immediately after hatching. In Spring it feeds on violet leaves. The magnificent Purple Emperor is basically black and white but the male has a lovely purple sheen when seen at the proper angle. The butterfly's feeding habits are, however, less beautiful – it is very fond of rotting animal corpses, fresh manure or stagnant puddles. This butterfly seems to be increasing on Box Hill and can occasionally be seen in woodland rides.

Fieldwork and follow up ideas.

Two possible projects are given here, the first of which concerns the use of colours for protection and mate-finding, and the second the use of flowers by butterflies. Both can be used to develop observational skills and can be further developed in follow up work, the first in considering reasons for camouflage or warning colours and the construction of foodwebs; the second in considering pollination mechanisms including those of orchids.

1. Butterfly, moth and minibeast colours.
Colours can be divided into two main kinds:

Protective colours – which are used to protect the animal from predators which might eat it. Protective colours can be of several types:-

(a) Camouflage colours where the animal may blend in with its background e.g. the Engrailed moth and the Mottled and Willow Beauty moths resting on tree trunks; Click beetles looking like plant buds.

(b) Protective resemblance where the animal resembles something which the predator would not eat e.g. a dead leaf – the Angle Shades moth; a dead twig – stick insects and Peppered moth caterpillars; even a bird dropping – the Alder moth caterpillar and Chinese Character moth.

(c) Warning colours are bright colours which advertise that the animal stings or tastes nasty e.g. ladybirds, wasps and Cinnabar moth caterpillars. Some edible animals also have these warning colours and this is called "mimicry" e.g. hoverflies (edible) mimic wasps (inedible).

(d) Flash colours are hidden until the animal is disturbed when it suddenly shows warning colours, e.g. the Eyed Hawk-moth.

Display colours

These are used to attract mates. The male butterfly is sometimes much more brightly coloured than the female e.g. the Common Blue or the Orange Tip, but in other species they are the same.

An important point here is that butterflies see a range of colours which is different from that which we see; they can recognise wavelengths from ultraviolet to orange, whereas we can see those between violet and red. Most protective colours are designed to be effective against the eyes of birds which perceive similar colours to humans, whereas display colours may appear quite different, to a butterfly's eye. For example the Dingy Skipper, dull brown to us, has a strong ultraviolet pattern.

Sit and observe butterflies to see whether butterflies settle where they are best hidden (this may not be true when they are feeding, but should be true when they are at rest). The children can sit and watch butterflies and record species, settling site, and colour/pattern of settling site.

Butterflies recognise potential mates by colour only, so watch to see if males persistently chase another butterfly of the same colour.

2. The use of nectar plants by butterflies.
Choose a sunny site with as wide a range of flowers as possible – perhaps a grassy patch with brambles close by. Watch a number of common flowers and record (a) which flowers are visited and (b) which butterflies visit which flowers and (c) how many butterflies of each kind visit each flower. Flower species likely to give good results include: wild marjoram, small scabious, brambles, rough hawkbit, knapweed, viper's bugloss, stemless and carline thistles, bird's-foot trefoil, bugle, wild thyme, fragrant and pyramidal orchids.

Suggested recording sheet:-

Recording Sheet

Butterflies

	Plant							
Butterfly	Bramble	Marjoram	etc					
Meadow Brown								
etc								

References:
British caterpillars by G. Hyde (Jarrold).
British moths by G. Hyde (Jarrold).
British butterflies by G. Hyde (Jarrold).
A field guide to the butterflies of Britain and Europe by L. Higgins & B. Riley (Collins).
The moths of the British Isles (I & II) by R. South (Warne).

MINIBEASTS

Minibeasts include insects and other small animals which are all invertebrates (have no backbones). There are more species of invertebrates than there are of vertebrates (such as mammals and birds).

Minibeasts live just about everywhere; they eat almost everything between them but are themselves eaten by a wide variety of other animals. Many minibeasts have evolved different kinds of adaptations as protection against their predators e.g. camouflage and behaviour. A great deal of information about minibeasts can be discovered by catching them, studying them and then of course letting them go.

Minibeasts likely to be found on Box Hill
ANNELIDS: The simplest minibeasts are worms. True worms are segmented (annelid is derived from the Latin word meaning rings or segments). They are very valuable animals and their activities in the soil, aerating and mixing it – 'ploughing' – are vital for the maintenance of its fertility.

Slugs and snails are MOLLUSCS. They have soft bodies, but the majority have some kind of shell to protect them. They glide along on a flat fleshy foot, lubricated by lots of slime. Most snails are hermaphrodite (male and female together). The largest land snail in the world is the African Giant snail, up to 25cms long. The largest land snail in Britain is the Roman Snail, up to 10cms long. It is a rare species nationally but is common on parts of Box Hill. Please do not disturb them.

CENTIPEDES are voracious carnivores and have a large pair of poison claws to help them catch their prey. They rarely have 100 legs, but always have one pair of legs per segment. The largest centipede in the world is the Giant Scolopendra from South America which grows to 26cms long! The largest British centipede is only 7cms long.

There are about 1 million living species of INSECT in the world and the class is divided into 29 orders, according to differences in their life histories and wing structures. The largest order is that of the beetles (Coleoptera) which contains about 300,000 species. A typical insect has 3 main parts to its body – head (with eyes, antennae (feelers), palps (taste organs), and mouthparts); thorax (with 3 pairs of legs and possibly one or two pairs of wings); and abdomen. Most insects change (metamorphose) during their lives. The greatest changes are undergone by butterflies and moths, flies, beetles etc. The young insect emerges from an EGG. At this young state it looks nothing like the adult and is called a LARVA. It then has to PUPATE during which stage it develops into an ADULT. Dragonflies, grasshoppers and some others do not pupate. The young insect called a NYMPH sheds its skin every so often and gradually grows into an adult. The most dangerous insect in the world is the common housefly which can transmit up to 30 diseases and parasitic worms to humans.

Roman Snail

Emperor Dragonfly

Woodlice, like all CRUSTACEANS, are very dependent on moisture and are therefore never found in really dry situations. They mainly feed on rotting material. They have 7 pairs of legs and roll up into a ball when disturbed.

MILLIPEDES are also found in damp situations and can often be confused with woodlice. However they can be distinguished by counting the legs; pill millipedes have 2 pairs of legs on nearly every segment. They usually eat living plants.

ARACHNIDS have 4 pairs of legs and large eyes. They do not have antennae, wings or three parts to their bodies. They include spiders (with bodies divided into 2 parts) and harvestmen (bodies not divided and the second pair of legs always the longest). Spiders are probably the most feared group of minibeast that children are likely to encounter. The largest spider in the world is the Bird-eating spider of South America – its body is 8 cms long and it has a leg span of 25cms. None of these have ever been found on Box Hill – yet!

Where do minibeasts live?

At least half of Britains 21,000 insect species depend upon deciduous woodland – not surprising when one considers that, but for man's relatively recent clearance, most of Britain would be covered in such forest. Native deciduous woodland is able to support many more insect species than introduced or coniferous plantations because (a) native British insects have had time to adapt to it and (b) there are more places to live (niches). A typical deciduous wood can be divided into 4 layers, Tree canopy, Shrub layer, Field/Herb layer, Ground layer. Also there is one other woodland habitat which is important for minibeasts – dead wood. Each habitat contains different minibeasts with different requirements, for example:-

(1) Canopy layer: many minibeasts here are herbivores (plant eaters) such as caterpillars. There are also some carnivores (meat eaters) e.g. lacewings.

(2) Shrub layer: there are many different shrub species with different associated minibeasts including herbivores such as the caterpillars of the Brimstone and Purple Emperor and carnivores

such as ladybirds (seven and two spot).

(3) Field/Herb layer: the leaves and flowers of many plants are essential as food for many minibeasts e.g. Fritillary caterpillars and violets; figwort weevils and figwort.

(4) Ground layer: leaf litter is a rich habitat for many detrivores (detritus feeders) e.g. bristletails and springtails. Decaying and living plant tissues are food for woodlice and millipedes. Carnivores and predators such as spiders, harvestmen and beetles are abundant here – one of the largest is the Violet ground beetle (2cm long).

(5) Dead wood: This supports a number of minibeasts such as bark beetles which lay their eggs under the bark and woodlice which eat the rotting wood.

Not all minibeasts live in woods and many can be found on the scrub edges or on the open downland. To a minibeast, even such ground vegetation forms a woodland in miniature and like a real woodland, also has distinct layers, which are homes for different types of minibeast.

A typical deciduous wood has 4 layers:-

Tree canopy   Shrub layer   Field/Herb layers   Ground layer

Fieldwork and follow up ideas.

1. Catching and observing minibeasts to study. All minibeasts must be returned to the wild after observation. Flying insects are best caught with a conventional butterfly net. Insects and other animals in the grass and longer vegetation can be caught using a sweep net. 'Sweep' the net to and fro in front of you as you walk through the grass. Make 4 or 5 sweeps before examining

A sweep net

what you have caught, usually grasshoppers, crickets, bugs, beetles, spiders and flies. Minibeasts in the trees can be caught by using a "beating tray" (an upside down umbrella or piece of old sheeting will do). Hit or shake the branches sharply and the creatures will fall off into the tray

A beating tray

below. Also search the tree trunks for resting moths etc. Crawling and other ground-living minibeasts can be found by 'hand searching' under logs etc. (remember to leave the logs as you found them). Smaller creatures can be caught

using simple 'pooters'. This apparatus consists of a small bottle with 2 tubes through the cork. One tube is flexible and is used to catch the animals. The other tube has a piece of gauze over its lower end so that, when you suck, the animals do not

Pooters can be any size

come up into your mouth. Pitfall traps will catch animals crawling along the ground. Each trap is simply a jar or pot sunk into the ground with the rim flush with the surface. Each trap has a simple roof such as a piece of wood raised up on stones to allow enough headroom for the larger minibeasts to get in. This lid stops rain from entering and drowning the animals. Make sure you take up your traps when you have finished with them.

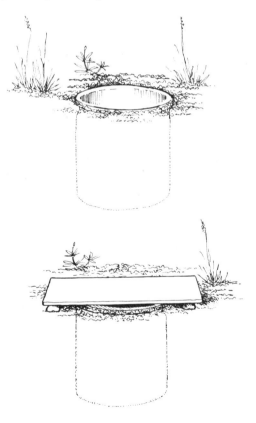

Pitfall traps

2.   Look at the minibeasts which live in leaf litter by using a Tullgren funnel.

The apparatus needs to be put together at school in advance of the sampling/collection. Collect a sample of leaf litter and sort through to remove any large invertebrates such as beetles, woodlice and spiders. These can be looked at separately. Put the sorted leaf litter into a sieve or netting and rest it in a funnel. Place a collecting pot under a funnel, position a 25 watt bulb near enough to illuminate and heat up the leaf litter (but take care that the dry litter does not catch fire). Leave the apparatus illuminated for at least 2 hours – preferably overnight. After this time examine the small invertebrates which will have fallen into the collecting pot whilst trying to escape the heat and light. Much more detail will be seen using a binocular miscroscope or magnifying glass. Use identification keys to help you identify these small invertebrates.

3.   How well are minibeasts camouflaged? Record the colour of the animal and the colour of the habitat in which it was found (see 'Butterflies and Moths').

4.   It is well worth looking at the classification of the invertebrates in some detail. Not all invertebrates are insects.

References:
Insects in Britain by G. Hyde (Jarrold).
Oxford book of invertebrates by D. Nichols & John Cooke (Oxford).
A field guide to the insects of Britain and N. Europe by M. Chinery (Collins).
The Oxford book of insects by J. Burton (Oxford).
The clue book of insects by G. Allen & J. Denslow (Oxford).

A Tullgren funnel

BIRDS AND MAMMALS

Remember that what you will actually see in terms of birds and mammals on Box Hill will depend on the time of year, the weather, the time of day and most importantly, how quietly you move and how carefully you look. Many birds and mammals appear seasonally on Box Hill and some are more apparent than others.

Spring and Summer give the greatest variety of birdlife, since many birds migrate South in Autumn to overwinter in warmer climates. Open grassland appears to support the least number of species, and the beech and yew woodland on Box Hill is so dark and shaded by dense canopy that bird sightings are rare. However, open woodland is home for many birds (especially where there is an abundance of undergrowth) and scrubland also supports a number of other species. All the common birds of the English countryside occur here, blackbirds, thrushes, robins, wrens, dunnocks, titmice and finches. Wood pigeons are a common sight, as are carrion crows and magpies. The brightly coloured jay is quite common and so too are starlings, mistle thrushes and in Winter, redwings. In the Summer, numbers increase as the migrants return – blackcaps, garden warblers, willow warblers, chiffchaffs and whitethroats. The cuckoo may be heard in Spring and the turtle dove adds its coo-cooing to the background chorus of birdsong. Woodland birds such as nuthatches, treecreepers and woodpeckers are occasionally seen, although it is necessary to walk very quietly to improve chances of a sighting. One bird that is seen and heard frequently is the pheasant. Common throughout England and Wales, these game birds were probably first introduced to Britain by the Normans in the late 11th century. Birds of prey are mainly confined to night-time flight although the kestrel is active during the day. The tawny owl is common in the woods all over Box Hill and there are several pairs of sparrowhawks around. Down at the river's edge, kingfishers, moorhens and herons may occassionally be seen.

If a cat-like tread and caution are necessary to see birds, then spotting mammals requires sitting patiently and quietly. Active predominantly in daylight hours are bank voles, commonest in early Autumn when the fruit and seed harvest has just finished. They may be seen scurrying amongst undergrowth at the edge of the woodland. Short-tailed voles, feeding mainly on grasses, may be heard squabbling with each other at the edge of arable land. If you are around at dusk or dawn, the larger mammals – rabbits, foxes, badgers (the largest land carnivore in Britain) and even roe

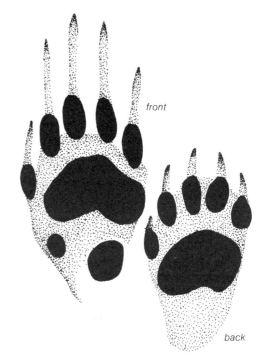

Badger footprints

deer can be spotted. The reduction in the traditional rural habitat has led to a decline in badger populations. However the "cunning" fox has managed to adapt and now lives happily in urban areas. You may see one basking on a roof, or find the evidence of a raided dustbin. More nocturnal species of mammal, which include woodmice and yellow-necked mice are common here; in fact the woodmouse is abundant in many parts of the woods. A census of the woodmouse population on Lodge Hill identified 43 individuals within a 0.81 hectares plot of woodland in June 1985. To see these creatures active you would have to sit quietly in a comfortable spot close to a known entrance tunnel or runway. Moles, bats, stoats and weasels do occur around Box Hill though are not as evident as the all-too-familiar grey squirrel. Introduced in the 1870s from North America this active rodent, which has few predators, can cause real harm to trees in the woodland by stripping rings of bark off trees. This destroys the phloem (food transporting tubes) and the roots are starved, killing the whole tree. Roe deer can also damage young trees and evidence of their browsing is clearest along woodland edges.

Fieldwork and follow up ideas.

It would be impractical to set out with a large class of pupils with the intention of bird or mammal spotting. The noise of feet, coughs and sneezes alert these animals to your presence. Look for signs instead; a chance sighting of a live creature will be a bonus.

1. Mammal tracks and bird footprints, once located, can be drawn, identified and even made into plastercasts (see mastersheet for instructions). Muddy paths can be the best places to search, and in winter when snow is on the ground numerous animal tracks can be seen very clearly.

2. Look for evidence of feeding. Empty nuts and seed cases may bear teeth marks; for example a hazel nut with a neat circular cavity in the outer case is evidence of bank voles or woodmice. Gnawed fir cones and bark stripped from trees are also evidence of other animals such as rabbits, deer and bank voles – look carefully at the height at which bark is gnawed and the actual marks left by the teeth.

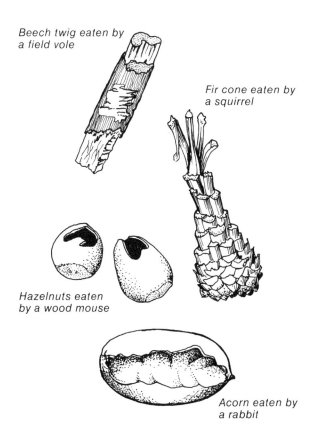

Beech twig eaten by a field vole

Fir cone eaten by a squirrel

Hazelnuts eaten by a wood mouse

Acorn eaten by a rabbit

Birds may partly eat fruits and seeds, so watch out for pecked hips, haws or crabapples. Often the seeds within the fruit are undigested and can be seen amongst bird droppings. Inspect holly leaves carefully and you might find V-shaped tears where a bluetit has extracted the larva of a holly leaf miner.

3. If you are in the area out of the nesting season and when the trees are losing their leaves try looking for birds' nests or animal holes – can you guess what might live in them?

4. Feathers, droppings or traces of fur caught on barbed wire or hawthorn bushes can be used to identify animals.

5. If you find any owl pellets examine their contents to give clues as to what they have eaten. Bones, teeth and fur are regurgitated in this undigested form and identification of the animal eaten can be made by using a simple key such as – 'Collecting and analysing bird pellets' by D. Glue, produced by the Young Ornithologists Club (YOC), The Lodge, Sandy, Beds. SG19 2DL.

6. Study any feathers that are found. Indentify the bird they have come from. Look at the way that feathers are made, how they work in flight and how they waterproof the bird.

7. Try making a bird's nest from dead twigs, dry grass, straw and mud. It's harder than you think!

8. Even keeping gerbils, white mice etc. at school can be a useful way of learning about behaviour, growth and adaptations of small mammals.

References:
Field guide to the birds of Britain and Europe by Peterson, Mountfort & Hollom (Collins).
Handbook of British mammals edited by G. B. Corbet & H. N. Southern (Blackwell).
Guide to animal tracks and signs by P. Bang & P. Dahlstrom (Collins).
Tracks and signs of British animals by A. Leutscher (Cleaver-Hume).

TREES

Trees are large plants and have a stem, leaves, flowers and fruit. Trees can grow to over 6m high on a single woody stem (shrubs are generally smaller and grow from several stems). Trees can be divided into two groups:- **conifers** and **broadleaves.**

Broadleaved trees have seeds enclosed in proper "fruits" like apples, pears and cherries. They tend to have wide flat leaves which they shed in autumn – "deciduous" (meaning "to fall down" in Latin). This prevents a loss of water when the roots cannot absorb much from the cold winter soil although holly and box are evergreen broadleaves. The timber from broadleaved trees is called hardwood and is used for making furniture.

Coniferous trees bear their seeds in cones; the large 'fir cones' are the female ones, the male ones containing the pollen are much smaller. However, yew and juniper berries are modified cones, with a juicy outer covering which attracts birds which eat them and so help dispersal. Most conifers have tough, narrow, needle-like leaves which they keep throughout the winter – "evergreen". However, the larch is a deciduous conifer. The timber from coniferous trees is called softwood and is used for building or making paper.

Not only are trees plants, but they are also the largest living things in the world. The largest tree in the world is a sequoia (a conifer) – called General Sherman. It is 83m high and estimated to weigh 2030 tonnes, which is 20 times the weight of an average blue whale! The common oak which is the symbol of the National Trust is a dwarf in comparison, although we tend to think of it as a big tree, growing up to 40m high. However, much of a tree is dead material. For example, the living part of the trunk is a very thin layer of tubes just below the bark. This means that the middle of the trunk can rot away leaving a hollow tree which can survive as long as the crucial outer section is healthy.

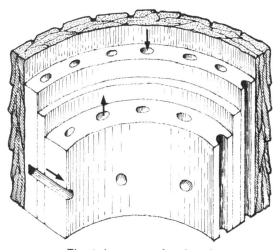

The tubes carry food and water around the tree

Trees are normally the final stage in the successional sequence (see 'Succession') although only a small proportion of our woodlands have evolved in this way. In fact, in Britain there is no **natural** woodland in the true sense of the word as all our woods have been altered and used by man to some extent. It is therefore more usual to refer to our woods as 'semi-natural'. The woodlands on Box Hill are thus largely the outcome of man's interference although the soil type has an important influence on the trees which grow on it. Chalk favours beech, yew and whitebeam, whereas the clay-with-flints favours oak, holly and birch although beech and ash also grow well on it. Competition between trees, in an area, is largely a matter of shade-creating and shade-tolerance.

Once the BEECH is established it casts a very heavy shade and suppresses most other forms of trees except box and yew which are shade-tolerant. The floor of an established beech wood is also almost devoid of small plants for there is too little light for ordinary green plants to grow. Beech leaves are also slow to decay (see 'On the Ground'). Beech has a very characteristic fruit; beech 'mast' is a bristly cup containing two or three triangular nuts, which are edible. They are a source of oil which was made into a kind of margarine in Germany in the two World Wars. The timber is mainly used for making furniture.

Beech 'mast' and nuts

After the beech, the YEW is the most common tree on Box Hill. It can tolerate the shade of the beeches and grows slowly to a great age. In many parts of the area it occurs as a solitary tree (e.g. on the Long Spur) but in Juniper Bottom and on The Whites there are so many yew trees that they have suppressed all other forms of vegetation and the ground beneath them is quite bare. Yew trees have been planted in churchyards for many centuries – the evergreen foliage being seen as a symbol of immortality. The timber is hard and was used to make archery bows but is now mainly used for turning, carving and veneers.

Yew 'berries' are poisonous

The OAK does not grow well on chalk and is mainly found on the clay-with-flints. There are four types of oak tree on Box Hill, the common or pedunculate oak with stalked acorns, the turkey oak, the holm oak (which is evergreen) and the red oak. The red oak is a native of North America; it has saw-edged leaves which turn a lovely red in the Autumn. Oaks grow very slowly but seldom reach any size on the Hill for when they are still relatively young (80-100 years old) their roots reach the Chalk underneath the soil which inhibits them. Oak trees do not cast a heavy shade like beech and yew, permitting a vigorous undergrowth of hazel, holly, bramble etc. to flourish; all provide shelter and food for birds and animals. More species of insects feed on oak than on any other kind of British tree. More than 100 species of moth have caterpillars that feed on oak leaves and they can sometimes strip a tree completely and a large oak tree can have as many as 250,000 leaves. The timber from oak trees was used for shipbuilding and the trees were often grown in such a way as to form bends or 'knees' as required by the shipwrights. It took 3,000 large oaks to build one man o'war for Nelson's navy.

Box Hill possesses one of the few semi-natural BOX woodlands in the country. Box is a small evergreen shade-tolerant tree which is frequently planted as garden hedges. It is a Mediterranean tree; growing wild here in the south of England it is at the very limit of its 'range'. The wood is hard and is used for making tool handles, chess-pieces and musical instruments.

The JUNIPER tree, which has given its name to so many local places, is another small evergreen tree but one which is very susceptible to shade. It seems to act as a 'nurse', protecting other species until they grow so high as to block out the light. When it dies it does not rot, but is preserved by the oil it contains. Many dead juniper trees can be seen in Juniper Bottom under the shade of the yew trees which killed them. These should not be disturbed under any circumstances. The berries of juniper trees are used for flavouring gin – the name 'gin' comes from "genévrier", the French for juniper.

Juniper 'berries'

The woods on Box Hill are a good example of woodlands managed for amenity, landscape and nature conservation. National Trust policy on Box Hill is to maintain a tree population of varying ages and predominantly made up of those native broadleaved species which grow well on chalk or clay-with-flint based soils. Plantings of conifers

have occurred in the past and these are being extracted progressively and used as timber on the estate. Trees now being planted are mainly native British species such as oak, beech, wild cherry, ash and lime. The danger point in the life of any tree on Box Hill is in the early stages of its growth, when it is particularly vulnerable to attack from the grey squirrel, rabbits and deer (see 'Birds and Mammals'). Trees are in fact vulnerable up to the age of at least 40. It is possible to protect young saplings with a tree-guard but it is impossible to protect the bigger trees except by controlling the pest.

Do not forget to examine the bark of the trees carefully for plants that live there. You may discover some lichens. They are called "epiphytes" and get their nourishment from rain water and dust; they are just using the tree trunk as a convenient parking place and do no damage. Lichens consist of two sorts of organisms, a fungus and an alga (a very simple plant) growing together. The alga benefits from a sheltered environment provided by the fungus. The fungus benefits from the nutrients that the alga produces

Shield lichen

through photosynthesis. This mutual dependence is known as "symbiosis" (from the Greek meaning – 'living together'). Lichens grow best where there is plenty of rain or mist. They grow extremely slowly and are susceptible to impurities in the air and are therefore very good 'indicators' of air pollution.

The clumps of mosses that grow on trees and walls etc. are made up of individual plants with a single stem and a cluster of leaves. They do not have flowers but reproduce by spores. The spore-bearing stage is just a simple stem with a capsule containing the spores on the top. Moss plants cannot retain water well and they easily dry up, but a clump of them together acts like a sponge to hold water. Sphagnum moss (the main plant in bogs) can hold up to 20 times its own weight in water.

Moss

23

Fieldwork and follow up ideas.

Woodlands are one of the most accessible ecosystems for all ages to study and there is a very wide range of fieldwork that can be undertaken, depending on the age and ability of the group. The following are just some ideas.

1.   Identification of tree species. This can be very simple or quite precise depending on the group. Use background knowledge or simple keys to establish whether the tree is conifer or broadleaf, evergreen or deciduous. Alternatively, collect enough information about the tree to be able to identify it properly back at school. To do this you will need to collect information on (a) the leaf – simple or compound, opposite or alternate (b) the twigs and buds – opposite or alternate, the shape and colour; (c) the bark – smooth or rough, colour and texture, look at the pattern by doing a bark rubbing; (d) the flowers and fruit if visible.

2.   The age of the tree. This can be estimated by measuring (in cms) the girth (circumference) of the tree at 1.5m high. Divide the answer by 2.5 and this will give the approximate age of the tree. There are some exceptions to this – namely the horse chestnut and many coniferous trees. The age of coniferous trees can be estimated by counting the set of 'whorls' of branches. Add two to this number to estimate the age.

The most accurate way to find out a tree's age is to count the annual rings on a tree stump or large branch. Each ring represents one year's growth; the width of the ring shows the speed of growth – wide rings indicate rapid growth during that year. This can be linked into weather records if they are available.

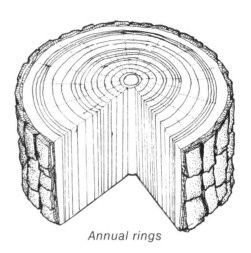

Annual rings

3.   Canopy cover. At least 8 people stand with their backs to the trunk and then walk outwards until they reach the edge of the tree – the last twig or leaf should be directly above their heads (make sure they are looking vertically). The distance around them can then be measured. The percentage canopy cover can be estimated by looking up at it through a plastic drainpipe (4cm diameter) with crosswires.

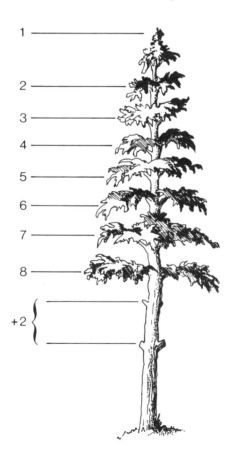

This conifer is 10 years old

Measuring the canopy

4. Height. This can be 'guesstimated' by comparing the tree with the height of a friend standing directly in front of it. A slightly more accurate way is to 'pretend to fell the tree' using a ruler.

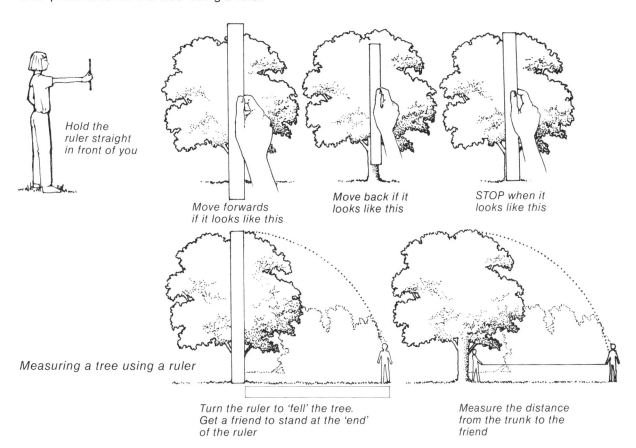

Hold the ruler straight in front of you

Move forwards if it looks like this

Move back if it looks like this

STOP when it looks like this

Measuring a tree using a ruler

Turn the ruler to 'fell' the tree. Get a friend to stand at the 'end' of the ruler

Measure the distance from the trunk to the friend

The most accurate way of measuring the height of a tree is to use a clinometer (see 'Geology' for instructions on making one). When the top of the tree is lined up with 45° the height is the horizontal distance from the tree to the person plus the vertical distance up to the person's eye.

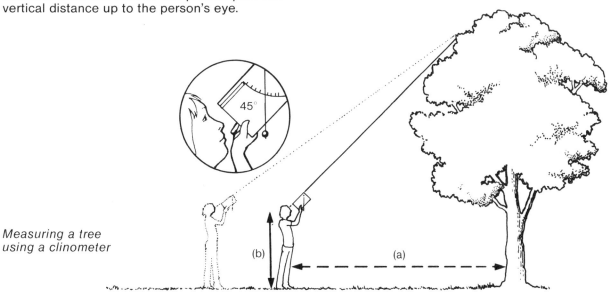

Measuring a tree using a clinometer

45°

(b)

(a)

When the clinometer is at 45° the height of the tree is (a) + (b)

25

5. A wood is more than a haphazard collection of trees and undergrowth. Look for the 'layers' (see 'Minibeasts') and compare those found under oak with beech and yew. What implications does this have for wildlife?

6. Look at growth rates of broadleaved and coniferous trees (contact the Forestry Commission at 231 Corstorphine Road, Edinburgh, EH12 7AT for these). Graph the results to see which is the faster, and therefore the more economical, to grow.

7. Leaf prints, leaf rubbings, bark rubbings, collages, mobiles, scrapbooks are all very visual ways of describing the tree. Comparisons can be made between different species.

8. A tree near the school can be 'adopted' and studied throughout the year.

9. Find out about the different uses for different types of wood. In the past, different types of wood were used for specific purposes e.g. elm is water-resistant and was used for making cart wheels and also coffins!

10. The leaves of many trees are often covered with 'galls' – abnormal growths caused by the activity of insects and mites. Collect galled leaves when they are turning brown and keep them damp in a plastic box. Watch for the insects to emerge in the spring.

11. Find out how slowly lichens grow – only a few mm per year, although they can live for several thousand years. Look at those growing on gravestones – they cannot be older than the date carved on it.

References:
The Family Naturalist by M. Chinery (MacDonald and Jane's).
A field guide to the trees of Britain & Northern Europe by A. Mitchell (Collins).
Trees in Britain by R. Randall (Jarrold).
Common British lichens by F. Dobson (Jarrold).

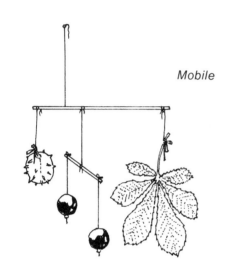

Mobile

Scrapbook

Project ideas

ON THE GROUND

A young child's world is down on the ground. A five-year-old out on a walk will watch the floor and pick up the treasures of feathers, twigs and stones which are to be found there. As we grow taller, we begin to look up more and often forget the place of interest to be found under our feet.

SOIL is the vital substance on the ground which supports life. A close look at a handful of soil will reveal it is not just "dirt" or, as one child described it, 'wet dust'. The first input of soil is broken-down rock (**mineral** or **inorganic** matter); into this, rotting vegetation and other **organic** matter is incorporated from the surface. The spaces between these particles are filled with **air** or **water**. A fifth component of soil is sometimes recognised, the organisms living in the soil and known as the **biota**.

Soils vary – different types of rock, known as the parent material, yield particles of mineral matter with certain characteristics, such as chemical composition, colour and size. These are important as they influence which plants are able to succeed in the soil. The chemical composition can determine the availability of vital nutrients while the size of the particles, or texture of the soil, determines the drainage of the soil. Large particles do not pack as closely as fine ones and thus allow freer drainage. This also affects the stability of the soil, a knowledge of which is crucial if construction of roads, houses, etc. is to be carried out.

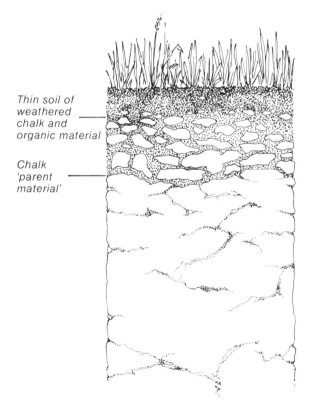

Thin soil of weathered chalk and organic material

Chalk 'parent material'

A Rendzina soil profile

The importance of soils to agriculture and to building to a lesser extent, has generated a lot of research and a complicated terminology. However it is possible for the youngest students to carry out simple investigations on soil (see Fieldwork and follow-up ideas). A comparison between the chalky soil known as a Rendzina with the soil on the clay-with-flints is very interesting. The calcium carbonate of the chalk parent material not only makes the soil alkaline, but also dissolves in rain-water. This means that most of the inorganic matter is washed away and very little is left to make up the soil, so only a very thin layer is built up. This can be seen at the top of the Burford Slopes, above The Whites river cliff. On the clay-with-flints a very different, thicker, acid soil builds up. This soil is also topped with a covering of leaves which fall from the many deciduous trees growing in the wood. These leaves and other organic matter are in various stages of decay.

Litter, fermenting & humus layers

The leaves on the surface making up the **litter layer** are from the last year's fall, and appear to be whole. However the usually fruitless search for a leaf without a hole in it shows that animals have already started the work of breaking down the leaves. Underneath the litter layer, it is still possible to identify remnants of leaves and twigs, but the detrivores have broken down the organic matter into much smaller particles. This layer is known as the **fermenting layer** and has a lovely rich smell. The organic matter is eventually decomposed by bacteria and fungi into an amorphous black **humus layer** which has little smell. It is at this stage that the organic matter is incorporated into the inorganic matter to make soil. The breakdown of organic matter is one of the most important processes for life on this planet. It occurs in fields, gardens, under hedges, etc., recycling the essential plant nutrients back into the soil. The layers mentioned are easy to identify in the Box Hill woods and are good for demonstrating how the process works. Leaf decay is slower under beech trees because the leaves are rich in tannins, chemicals which preserve the leaves and deter invertebrates. The thick foliage affects the plants which grow on the woodland floor since they require light to manufacture their own food. These are mainly herbs (plants which do not have woody stems and which grow and die down every year). Many grow early in Spring, to produce flowers and seeds before the canopy of tree foliage closes in on them. Carpets of bluebells can be seen in the Ashurst Rough woods in Spring; they die back transferring their food material to underground bulbs in the Summer.

The beechwoods on the Hill have little or no herb layer because of the lack of light. However organisms which do not manufacture their own food using sunlight **can** exist here. Fungi are classed as a kingdom separate from plants and animals. They feed off other plants and animals using fine thread-like cells called hyphae which grow into the food source. Some feed off dead material and act as decomposers. Others feed off living organisms – if this harms the plant or animal, the fungus is a parasite. An example of this, known to most children, is the fungal growth which develops on weak or injured goldfish. Birch polypore is another example which you may see

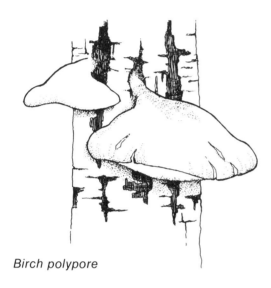

Birch polypore

growing on the birch trees on the clay soil. Not all fungi harm their 'host' (the plant or animal on which it is growing). The colourful red and white spotted fly agaric feeds on birch roots and the fungus and tree help each other: the tree roots provide sugar for the fungus which, in turn, passes on nutrients that it has broken down to the tree. This enables the tree to prosper in poor soils.

The hyphae are present throughout the year, but it is the fruiting bodies, often appearing in Autumn, which indicate the presence of fungi. These fruiting bodies disperse thousands of tiny spores for reproduction. There are 3,000 larger fungi in Britain and it takes a great deal of skill and practice to identify them. However the layman can see how the fruiting bodies vary because of the different methods the fungi use to disperse their spores. The gills we see on the underneath of a mushroom are used to drop the spores into air currents so that they are blown to new locations where they can start to develop. In other fungi the underneath has a spongy texture made up of hundreds of tubes from which the spores are dropped. The 'caps' of gilled fungi (known as agarics) and the tubed fungi (known as boletes) are on stems so the spores drop from a height into the air currents. Other fungi achieve height by growing on tree trunks and these are called bracket fungi. Some fungi grow on the ground

and thus have to shoot their spores up into the layer of moving air. You may be lucky enough to find an earthstar fungus in the beech woods. The outer layers of this onion-shaped fungus peel back which makes it look like a star. The central body disperses the spores when raindrops make impact. Some people would consider themselves unlucky if they came across a stinkhorn; this phallic fungus produces a slime smelling of rotting flesh to attract flies which, by feeding on the slime disperse the spores contained in it.

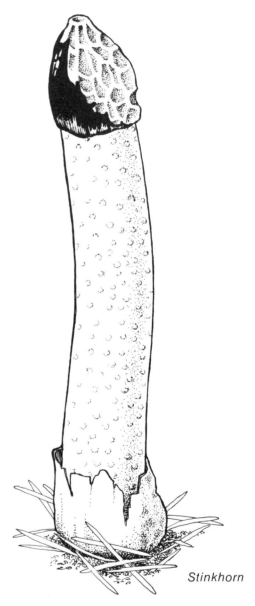

Stinkhorn

It is only possible to mention a few of the fungi here. Several reasonably priced books will help identify the more common ones; two are listed at the end. Fungi are a fascinating group with colourful, sometimes rather grisly names which appeal to children. The fact that some fungi are extremely tasty also adds incentive to many a 'fungus foray'. However some fungi are poisonous, even lethal, and it is best not to eat or taste any fungus unless an expert has identified it.

Fieldwork and follow up ideas.

1. LEAF LITTER.

    (a) Examine the layers at different stages of decay by sight, feel and smell. Consider the importance of the first stage of breakdown of the leaves, which increases surface area. Simple reaction experiments can show the importance of this, back in the classroom. Adding hydrogen peroxide to a cube of raw potato will produce much less oxygen, than if the chemical is added to the same amount of raw potato which has been cut or mashed up. This experiment should be carried out taking appropriate precautions.

    (b) Minibeasts. Look for the minibeasts which live in this important layer. Hand searching will reveal a good sample and even more will appear if a simple Tullgren funnel is constructed (see 'Minibeasts').

2. FUNGI. Fungi have already distributed vast numbers of spores by the time the fruiting body has developed. Collection, therefore, does not threaten future populations. However, the National Trust's permission is needed before collecting anything on Box Hill but wherever you are it is a good idea to take only one of each kind and to avoid disturbing rare species. If you find cap fungi or buy some fresh field mushrooms you can make a spore print. Cut off the stalk as close to the under-surface as possible. Choose a piece of paper, the colour of which contrasts with the underside of the fungus. Lay the cap on the paper and cover with a large bowl to protect it from draughts. Leave overnight and then carefully lift off the cap; the spores which have been dropped will be left as a delicate pattern which can be preserved if sprayed carefully with artists' fixative (used for chalk or charcoal drawings).

3. SOIL. Compare the soil on chalk with that on the clay-with-flints at the locations shown on the map.

    (a) Colour
    Make a **soil smear** by moistening some soil and rubbing it on a piece of white paper.

    (b) pH
    Measure the acidity or alkalinity of the soil using pH tape or simple chemicals (available from BDH Chemicals Ltd., Broom Road, Poole, Dorset. BH12 4NN). The pH is indicated by the colour of the tape or the solution in the test-tube which is compared to a chart supplied with the chemicals or tape. The chalk soil will give a lovely blue-green colour showing its alkaline status, whereas the clay produces an orange colour showing that it is acid.

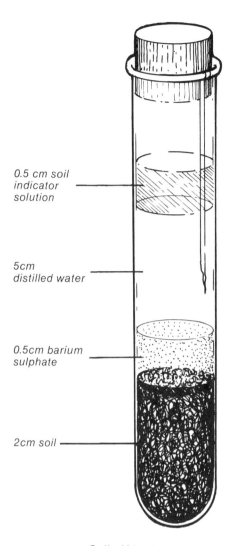

0.5 cm soil indicator solution

5cm distilled water

0.5cm barium sulphate

2cm soil

Soil pH test

SHAKE WELL AND LEAVE TO SETTLE OUT. Compare colour of solution with chart supplied with the chemicals.

(c) Texture
Texture relates to the size of the mineral particles in the soil. They fall into three groups, classed by the diameter of the particles.

Sand: 2mm – 0.06mm

Silt: 0.06mm – 0.002mm

Clay: Less than 0.002mm

Obviously, the particles are too small to measure with a ruler; texture can be assessed by 'feel' or by sedimentation (next page).

DIRTY HANDS TEST

Remember this test is designed to assess the size of mineral particles – a soil sample with lots of organic matter may not conform.

1.   Take a small handful of soil and remove all particles over 0.2mm in size. Moisten the soil – you may have to spit on it.

2.   Can you roll the soil into a ball and/or a sausage?

Yes – go to question 3.

No – the soil is mainly SAND.

3.   Wet the soil a little more. Can you roll it into a finer thread about 5mm thick and bend it into a horseshoe shape?

Yes – go to question 5.

No – go to question 4.

4.   Wet the soil again and rub it between your fingers – how does it feel?

Smooth – the soil is mainly SILT.

Gritty – the soil is SANDY SILT loam.

5.   Can you make a small ring by joining the ends of the horseshoe?

Yes – the soil is mainly CLAY.

No – the soil is a LOAM with CLAY.

SEDIMENTATION

Different sized particles will separate out in water because of different weights. Put dry soil in a jar of water and stir or shake.
Leave to settle out.

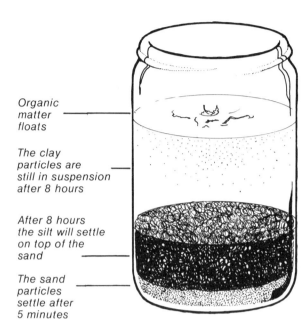

Organic matter floats

The clay particles are still in suspension after 8 hours

After 8 hours the silt will settle on top of the sand

The sand particles settle after 5 minutes

Sedimentation

References:
Spotter's guide to mushrooms & other fungi by R. Clarke (Usborne pocketbooks).
Mushrooms and toadstools by J. Wilkinson & S. Buczacki (Collins Gem series).
Ground Work – Practical Ways of Learning about Soils by M. Jarman (Leicestershire Museums Publications No. 53).

PUBLIC PRESSURE

Over half a million people visit Box Hill every year, to walk and picnic. On hot days they bask in the sun, while in the winter, given enough snow, the slopes are full of skiers and sledgers.

The use of Box Hill for such activities is no new thing. People have been coming for a long time to enjoy the fresh air and beautiful surroundings. In Jane Austen's 'Emma' a picnic at Box Hill is arranged. Unfortunately none of the scenery is described but it shows how the Hill was a place for 'gentlefolk' in the early 1800s. The opening of the railway (see 'Human Pattern') enabled the less wealthy and those further afield to reach the area. It became the destination for South Londoners' day trips. Walter Miles wrote in *Footpath Rambles Round Dorking* published in 1903 that Box Hill was "... a charming resort, beloved alike by the modest pedestrian and cyclist, as well as those who come down from London in style in a four-in-hand".

The early visitors were the guests of private landowners who were generous enough to allow the trippers to enjoy Box Hill. There may have been some extra revenue to be made from ventures such as the fair held on Donkey Green with stalls, shows and as the name implies, donkey rides. During this century, the National Trust has become the owner of Box Hill and much of the land around it. The estate now comprises of around 486 hectares (1200 acres) which is managed to achieve the charity's aim of preserving places of historic interest and natural beauty for

the nation. The first step of the consolidation of the estate occurred in 1914 when Leopold Salomons presented 93 hectares (230 acres) to the Trust which he had purchased in 1913 for £16,000. This included the summit where the Salomons Memorial now presides over the viewpoint. Further gifts and purchases followed, some only a hectare or two and others quite substantial holdings: Lodge Hill, donated in 1921; the woods of Ashurst Rough and Flint Hill comprising 100 hectares, bought in 1923; Mickleham Downs, donated in 1939; the Juniper Hall estate purchased in 1949.

Public pressure on this area is an interesting topic with much scope for investigations on where visitors come from and which parts of the Hill they favour (see Fieldwork and follow up ideas). The information gained from such surveys not only satisfies curiosity but is also useful to the National Trust which has the job of managing Box Hill. Positive management is necessary to maintain the character of the hill; two examples of this are the programme of sheep grazing implemented to prevent scrub encroachment (see 'Succession') and the construction of steps to reduce wear and tear on paths. There is also the need to reconcile the sometimes conflicting roles of the hill – on the one hand, a sanctuary for rare species and on the other, a playground for the not so rare. In 1971, Box Hill was the first Country Park to be designated by the Countryside Commission. This confirmed its importance as a place for outdoor recreation. The value of the flora and fauna of the hill is also recognised by the Nature Conservancy Council's classification of the area as a Grade 1 Site of Special Scientific Interest (SSSI).

The conflict can arise when public pressure leads to a decline or change in the plant and animal species. This can threaten already rare or endangered species e.g. autumn lady's tresses and bee orchids, Chalkhill and Adonis Blue butterflies. Certain activities may also detract from the visual appeal of the area, for example, eroded paths and litter. It is probably true to say that most of the visitors are unaware of the rarer species on the Hill which merit its SSSI grading. However surveys have shown that they appreciate and value the sum of all these species, be they rare or common. They come for the 'wildlife', the 'landscape' and the 'open air' which may be missing from their own environment. Without proper

Adonis Blue

Chalkhill Blue

management, there is always the danger that public pressure can actually destroy the very things that have attracted people there in the first place. In the case of Box Hill the 'first place' was about 300 years ago – today we are responsible for handing on this resource to the generations of the next centuries.

Lady's tresses

Bee orchid

CONSERVATION
This pack has only briefly covered the complex nature of Box Hill. Ironic as it may be, virtually every feature, results from the intervention (intentional or otherwise) by man. As man's circumstances have changed, so these features have become threatened and without a deliberate policy to protect and conserve them, they are bound to disappear over time. The threats come from two major sources:

(1) From man; by processes of industrial and urban expansion, agricultural change, forestry and recreation.

(2) From nature itself: no more than the natural course of succession. National Trust policy is to conserve the greatest variety of habitats by maintaining the different stages of succession. This is not easy and requires 'active' management. Downland is conserved by cutting back the encroaching scrub, followed by heavy Winter grazing by sheep from neighbouring farms. The sheep are usually upland or cross-breeds as

these seem to do better on the coarse grass than lowland breeds. Grazing during other seasons is then introduced as the downland improves. Some areas of downland are enclosed by permanent fencing (with stiles for access), whilst other areas are only temporarily enclosed by electric fencing which can be moved. The later stages of succession are allowed to develop elsewhere on the Hill. Large areas of scrub are maintained providing a valuable habitat. Woods are managed to achieve a mixed age range for landscape, amenity and nature conservation reasons. Some of the fallen trees are left for fungi and minibeasts.

Fieldwork and follow up ideas.

1. Origin of Visitors. Survey the car number plates in the car-parks. The *index mark* is the last two letters in the three letter combination and is specific to the Local Vehicle Licensing Office (LVLO) where the car was registered. A count of cars from nearby LVLOs will give an idea of where the visitors to Box Hill are coming from. A flow map is good for illustrating the results. Lines are drawn connecting the LVLO to Box Hill. The width of the line corresponds to the number of cars. Separate groups of children can look for plates from one office. The mastersheet includes the various codes for nearby LVLOs and an outline for the flow map. A complete list of index marks and licensing offices can be found in the HMSO publication V382 *Vehicle Registration and Licensing.*

2. Effect of Public Pressure.

(a) Trampling Survey: (see 'Grasses and other Plants').

(b) Pressure: Back in the classroom look at the effect of standing on a polystyrene tile in trainers and stiletto heels. Work out the pressure (kg per square centimetre) for the shoes and then compare these figures to one calculated for horses. Think how the pressure varies with different strides – when a horse is walking the weight is distributed between three legs for most of the time but when galloping there is a stage when all the weight is concentrated on just one leg. This impact together, with the shearing action of the horseshoe churns the ground up – the group may spot 'No Horses' signs to prevent this damage occuring in some areas.

3. Visitor Behaviour. Visitors to the Hill tend to cluster around certain key points – the viewpoint, Donkey Green near the car park and the cafe/restaurant/shop area. Locations that attract the visitor in this way are called 'honeypots'. Interesting surveys can be carried out to see how numbers drop off away from these honeypots.

(a) Matchstick measure.
At the beginning of the day insert a line of dead matchsticks across various paths (suggested points shown on the map). The matchsticks should be not more than 1cm above the surface. At the end of the day it is possible to see which paths have been used the most by the proportion of matches that have been broken.

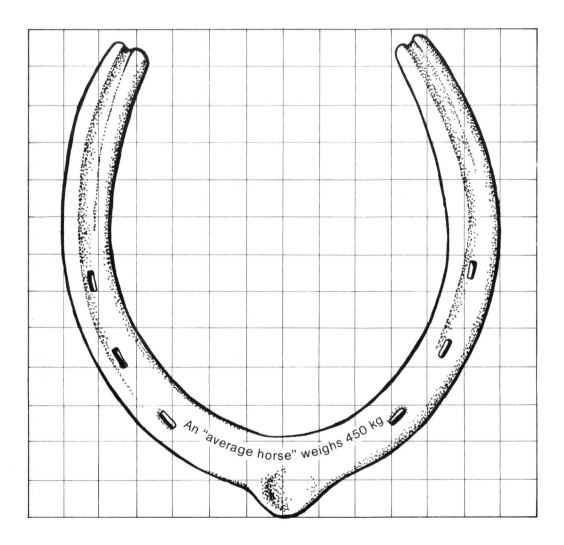

An "average horse" weighs 450 kg

(b) Transect.
Choose several routes radiating out from one of the honeypots (suggestions again shown on the map). The children pace out set intervals and note the number of people and the activity of those people on a simple recording sheet.

The results for this survey look most attractive if presented as a bar graph.

Breakdown of Activity

Paces	People passed	playing	walking	picnicing	viewing
1-10	15	II	I	III	₩ IIII
11-20	8	₩ I		II	
21-30	4	II	II		

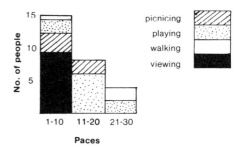

34

MASTERSHEET LANDSCAPE SKETCHING

Label the vertical lines with the correct descriptions:

Friends Provident	graveyard	St Martin's Church	Sewage works	A25	Railway Reading to Tonbridge	golf course	Boxhill Farm

Brockham	Chanctonbury Ring	Railway London to Horsham	Tower Hill	cricket pitch	Dorking	Leith Hill

Distant View

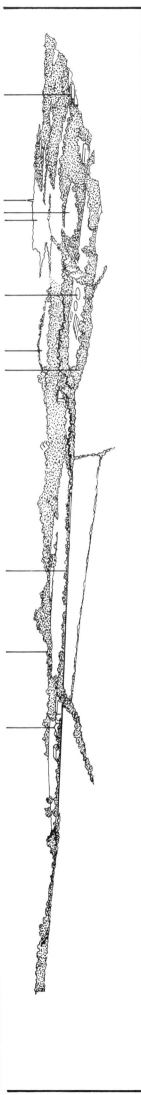

Middle-distant view

Shade these and cut them out neatly. Stick them
on top of each other to see the whole view.

Foreground

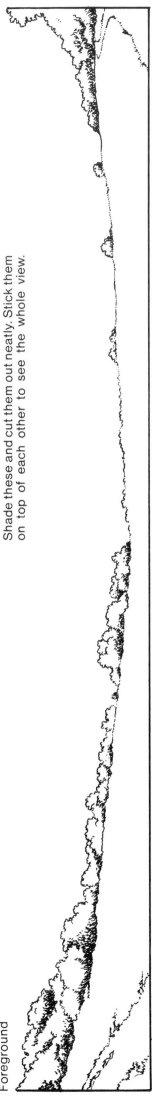

MASTERSHEET *GEOLOGY*

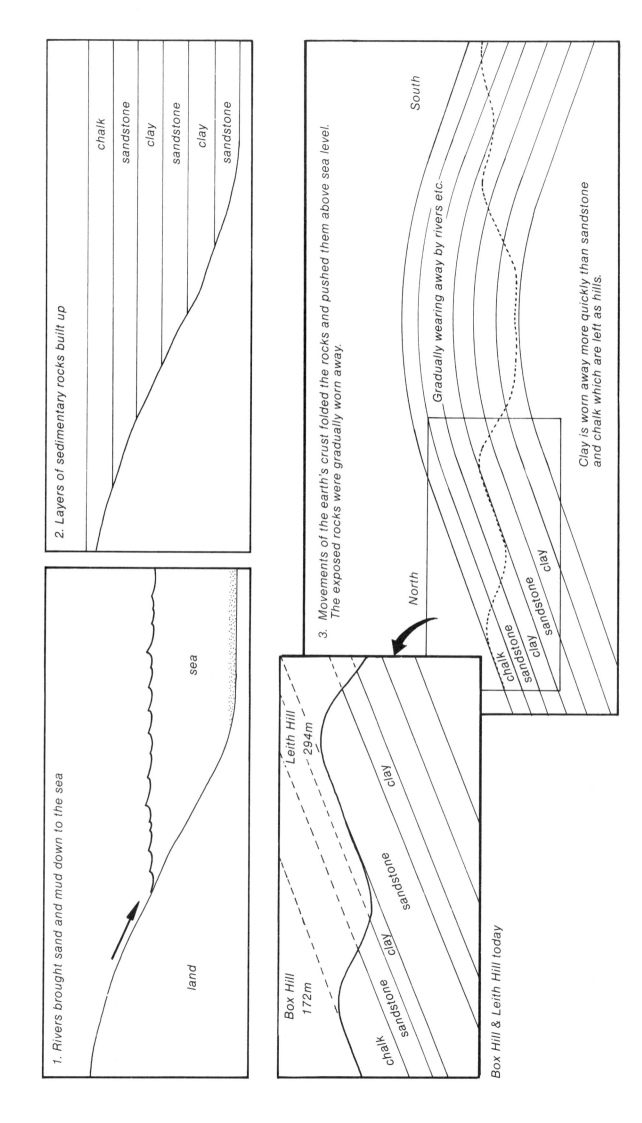

1. Rivers brought sand and mud down to the sea

land

sea

2. Layers of sedimentary rocks built up

chalk
sandstone
clay
sandstone
clay
sandstone

3. Movements of the earth's crust folded the rocks and pushed them above sea level. The exposed rocks were gradually worn away.

North

South

Gradually wearing away by rivers etc.

chalk
sandstone
clay
sandstone
clay

Clay is worn away more quickly than sandstone and chalk which are left as hills.

Box Hill
172m

Leith Hill
294m

chalk
sandstone
clay
sandstone
clay

Box Hill & Leith Hill today

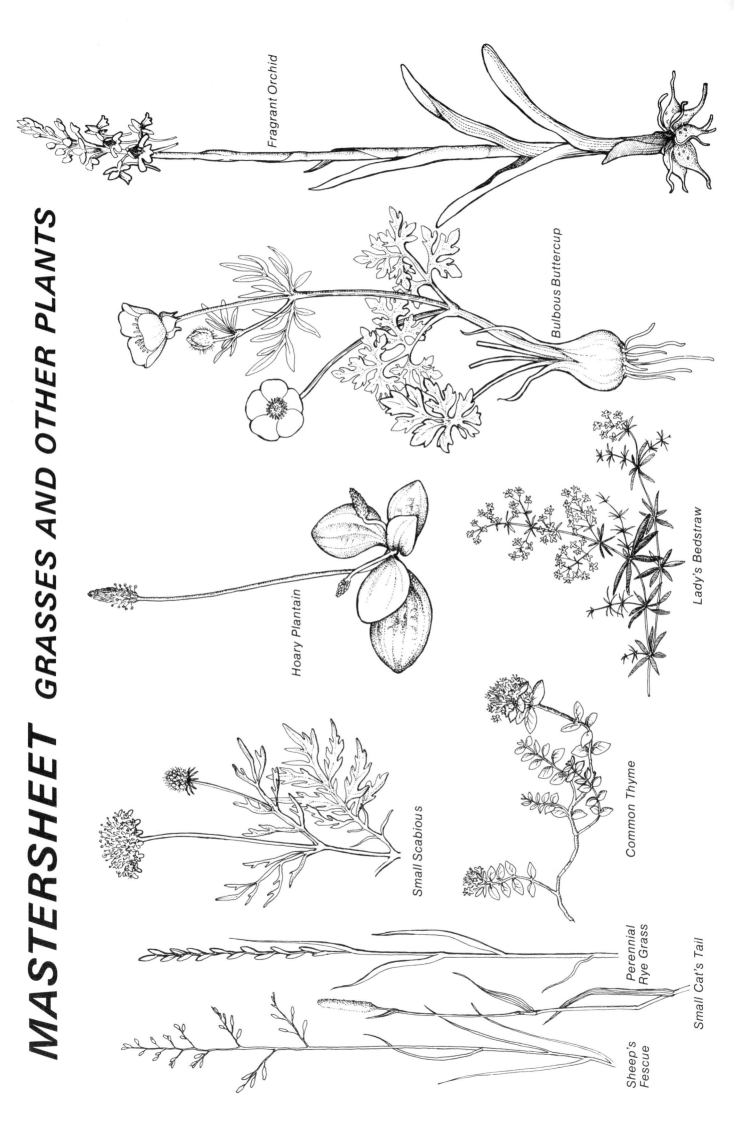

MASTERSHEET GRASSES AND OTHER PLANTS

Fragrant Orchid

Bulbous Buttercup

Hoary Plantain

Lady's Bedstraw

Small Scabious

Common Thyme

Perennial Rye Grass

Small Cat's Tail

Sheep's Fescue

MASTERSHEET BUTTERFLIES AND MOTHS

Small Copper

Peacock

Dingy Skipper

Comma

Burnet Moth

Brimstone

MASTERSHEET MINIBEASTS

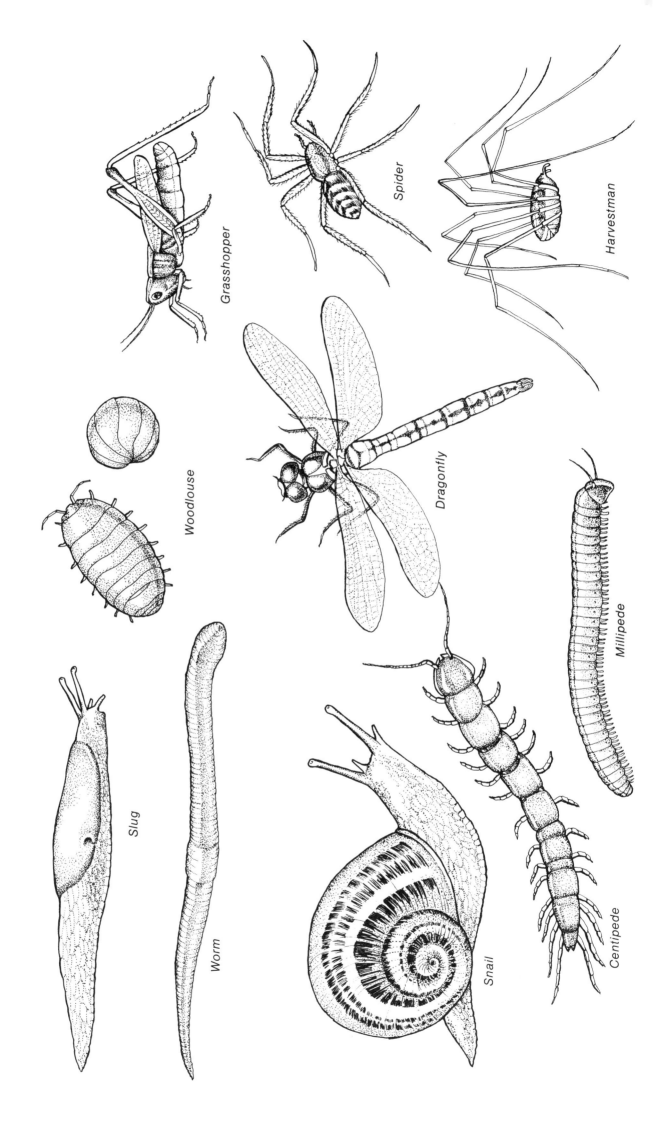

Grasshopper

Spider

Harvestman

Woodlouse

Dragonfly

Millipede

Slug

Worm

Snail

Centipede

MASTERSHEET *MAMMALS*

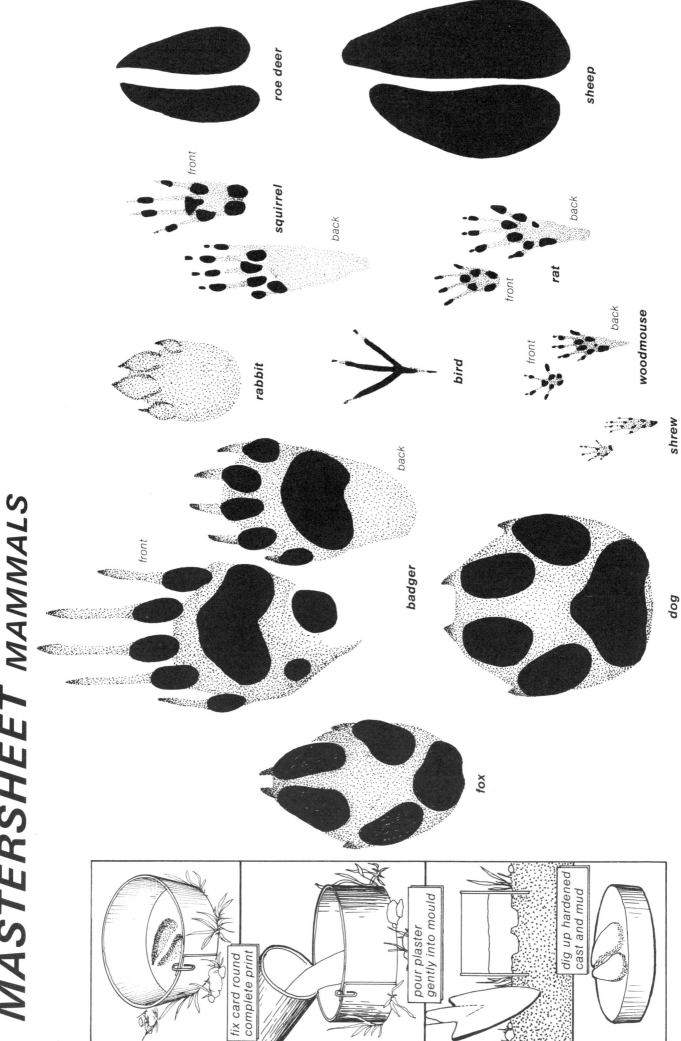

roe deer

sheep

squirrel — front — back

rat — back — front

rabbit

bird

woodmouse — front — back

shrew

badger — front — back

dog

fox

fix card round complete print

pour plaster gently into mould

dig up hardened cast and mud

MASTERSHEET TREES

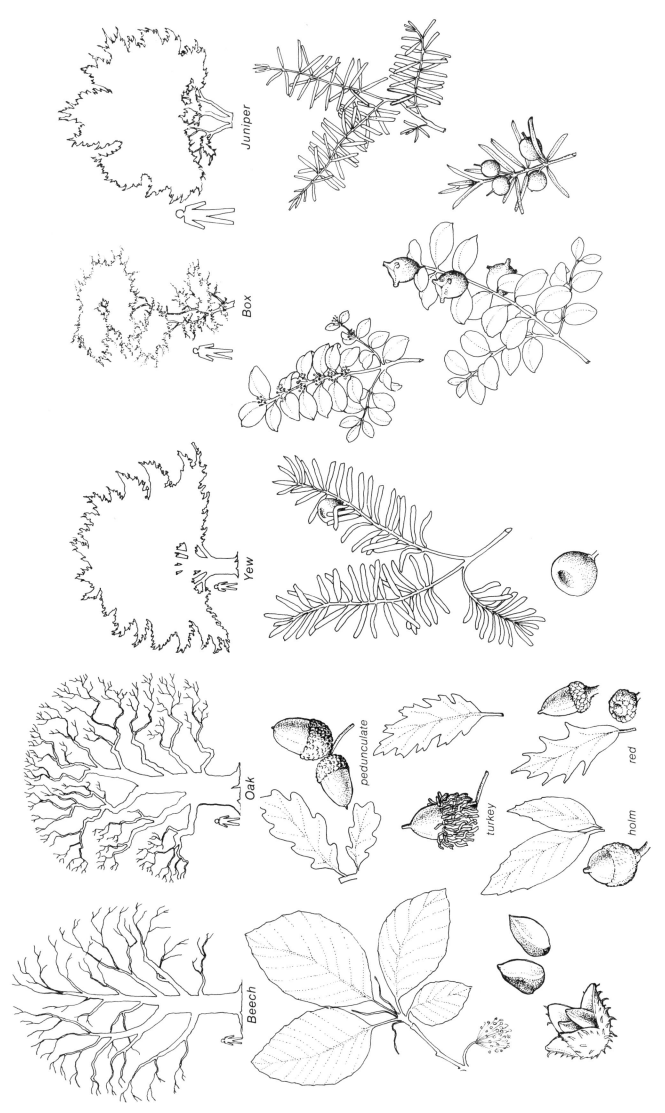

Juniper

Box

Yew

Oak
pedunculate
turkey
red
holm

Beech

MASTERSHEET *FUNGI*

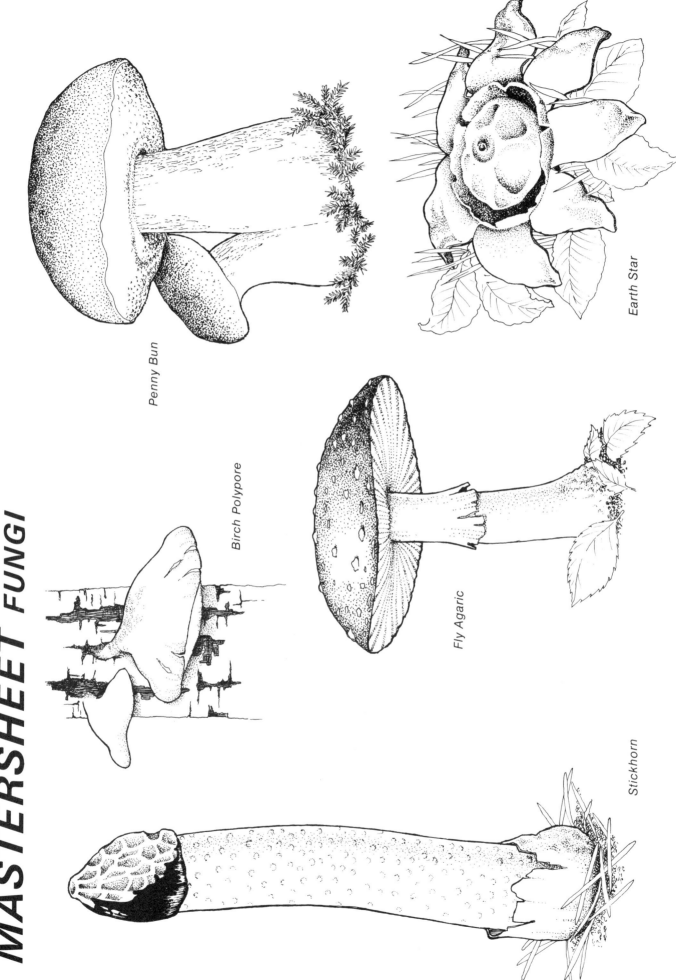

Penny Bun

Earth Star

Birch Polypore

Fly Agaric

Stickhorn

MASTERSHEET

WHERE DO THE VISITORS TO BOX HILL COME FROM?

The second two letters in the three letter combination of a British vehicle number plate tells you where the car has been registered.

B342 HEG   **AYB 363S**

The class is going to go round the cars in the car park and see what they can find out about the visitors to Box Hill from the number plates. **You** may have to look for letters that match those in one of the columns *or* count unusual number plates *or* count all the cars:- MAKE SURE YOU KNOW WHAT YOU ARE DOING.

Date: _____ Weather: _____

Time started: _____ Time finished: _____

Record using a Gate Tally

LONGON					Guildford
Central	Stratford	Ruislip	Merton	Sidcup	
HM,HV,HX,JD UC,UL,UU,UV UW,YE,YF,YH YK,YL,YM,YN YO,YP,YR,YT YU,YW,YX,YY	MC,MD ME,MF MG,MH MK,ML MM,MP MT,MU	BY,LA,LB,LC LD,LE,LF,LH LK,LL,LM,LN LO,LP,LR,LT LU,LW,LX,LY OY,RK	GC,GF GH,GJ GK,GN GO,GP GT	GU,GW GX,GY MV,MX MY	PA,PB PC,PD PE,PF PG,PH PJ,PK PL,PM

TOTAL Number of cars in Car Park

Number of unusual plates
(foreign, forces, personalised)

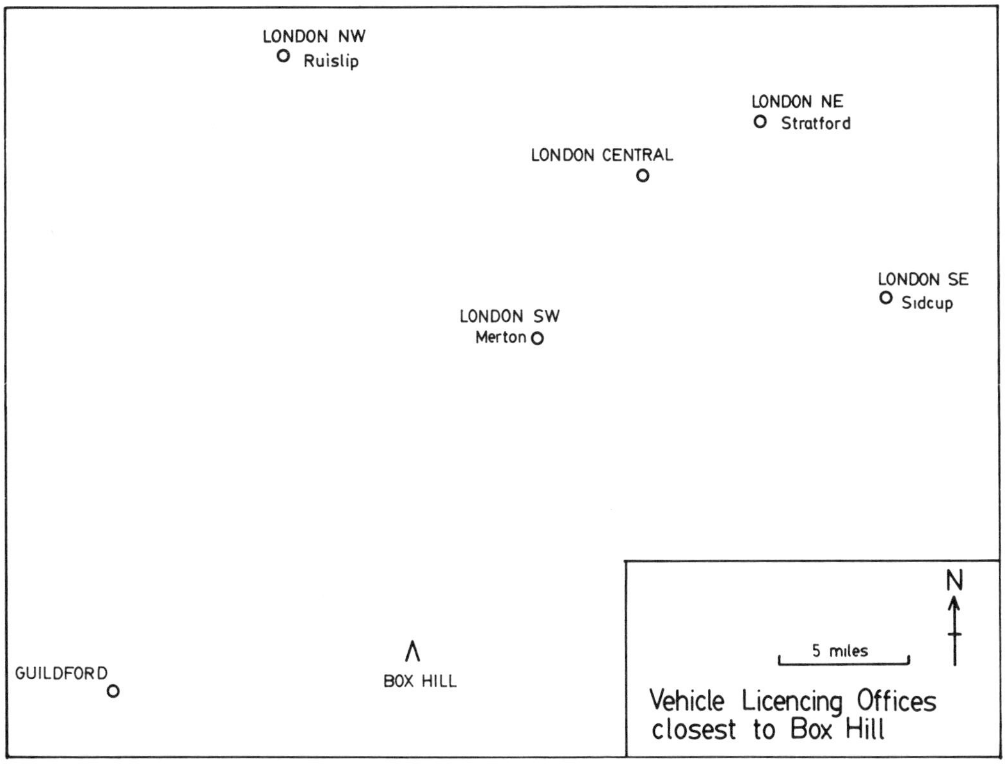

LONDON NW
O Ruislip

LONDON NE
O Stratford

LONDON CENTRAL
O

LONDON SE
O Sidcup

LONDON SW
Merton O

GUILDFORD
O

∧
BOX HILL

N
↑

⊢___ 5 miles ___⊣

Vehicle Licencing Offices
closest to Box Hill

BOX HILL

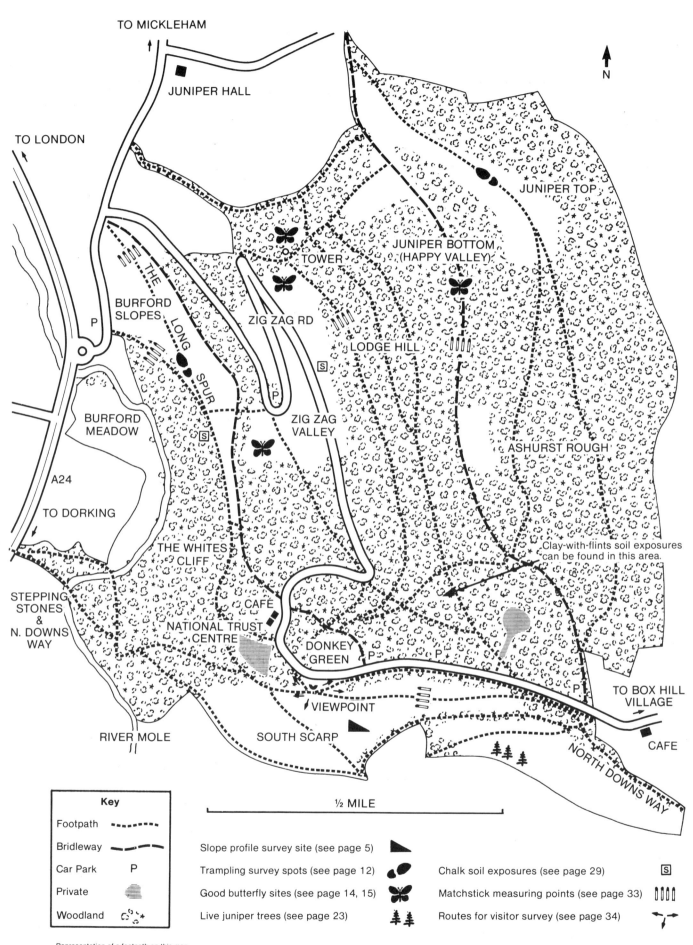

TO MICKLEHAM

JUNIPER HALL

TO LONDON

N

JUNIPER TOP

TOWER

JUNIPER BOTTOM
(HAPPY VALLEY)

THE LONG SPUR

BURFORD SLOPES

ZIG ZAG RD

LODGE HILL

ASHURST ROUGH

BURFORD MEADOW

ZIG ZAG VALLEY

A24

TO DORKING

Clay-with-flints soil exposures can be found in this area.

THE WHITES CLIFF

CAFÉ

NATIONAL TRUST CENTRE

DONKEY GREEN

STEPPING STONES & N. DOWNS WAY

TO BOX HILL VILLAGE

CAFE

VIEWPOINT

RIVER MOLE

SOUTH SCARP

NORTH DOWNS WAY

Key

Footpath	-------
Bridleway	— — —
Car Park	P
Private	
Woodland	

½ MILE

Slope profile survey site (see page 5)

Trampling survey spots (see page 12)

Good butterfly sites (see page 14, 15)

Live juniper trees (see page 23)

Chalk soil exposures (see page 29) — S

Matchstick measuring points (see page 33)

Routes for visitor survey (see page 34)

Representation of a footpath on this map does not necessarily imply a right of way.